Smart Innovation, Systems and Technologies

Volume 61

Series editors

Robert James Howlett, KES International, Shoreham-by-sea, UK
e-mail: rjhowlett@kesinternational.org

Lakhmi C. Jain, University of Canberra, Canberra, Australia;
Bournemouth University, UK;
KES International, UK
e-mails: jainlc2002@yahoo.co.uk; Lakhmi.Jain@canberra.edu.au

About this Series

The Smart Innovation, Systems and Technologies book series encompasses the topics of knowledge, intelligence, innovation and sustainability. The aim of the series is to make available a platform for the publication of books on all aspects of single and multi-disciplinary research on these themes in order to make the latest results available in a readily-accessible form. Volumes on interdisciplinary research combining two or more of these areas is particularly sought.

The series covers systems and paradigms that employ knowledge and intelligence in a broad sense. Its scope is systems having embedded knowledge and intelligence, which may be applied to the solution of world problems in industry, the environment and the community. It also focusses on the knowledge-transfer methodologies and innovation strategies employed to make this happen effectively. The combination of intelligent systems tools and a broad range of applications introduces a need for a synergy of disciplines from science, technology, business and the humanities. The series will include conference proceedings, edited collections, monographs, handbooks, reference books, and other relevant types of book in areas of science and technology where smart systems and technologies can offer innovative solutions.

High quality content is an essential feature for all book proposals accepted for the series. It is expected that editors of all accepted volumes will ensure that contributions are subjected to an appropriate level of reviewing process and adhere to KES quality principles.

More information about this series at http://www.springer.com/series/8767

Sergey V. Zykov

Crisis Management for Software Development and Knowledge Transfer

 Springer

Sergey V. Zykov
Higher School of Economics
National Research University
Moscow
Russia

ISSN 2190-3018   ISSN 2190-3026   (electronic)
Smart Innovation, Systems and Technologies
ISBN 978-3-319-82707-0   ISBN 978-3-319-42966-3   (eBook)
DOI 10.1007/978-3-319-42966-3

Printed on acid-free paper

This Springer imprint is published by Springer Nature
The registered company is Springer International Publishing AG Switzerland

To God, my teachers, and my family

Foreword

Based upon its title, "Crisis Management for Software Development and Knowledge Transfer," the reader might be forgiven for expecting that the subject of the book is addressing situations where development has gone disastrously wrong. Nothing could be further from the truth. Instead, Sergey Zykov has written an important book that offers value on many levels. For the student or new developer, it offers a useful survey of the important software engineering approaches that have emerged over the past six decades. For the more experienced developer or manager, it presents a thoughtful perspective on how each development methodology's approach to addressing crises can play in choosing the most appropriate methodology for a particular context. For the academic researcher, it proposes a number of interesting frameworks and identifies the central role that knowledge transfer plays in the software engineering process. In other words, crises cannot necessarily be avoided, but they can be more effectively managed.

By beginning the book with an examination of the software crisis of the late 1960s and early 1970s, Zykov performs an important service. A few decades ago, individuals that had lived through this tumultuous period were well represented in the typical IT staff. Today, most of that group is retired or deceased. Unfortunately, their absence has left a vacuum of experience. The anarchy that characterized mid-1960s software development has many parallels to much of the dot.com and Internet development of the past decade. Too often, we see history repeating itself in the form of systems that do meet user needs that fail to achieve acceptable levels of security or that vastly underperform expectations. Through better understanding the past, perhaps we will be less surprised by these outcomes. Perhaps we may even be able to avoid them.

Almost any reader can benefit from the book's systematic coverage of different software engineering methodologies. Emphasizing from the start that there is no "silver bullet" approach, Zykov clarifies often subtle differences between approaches—e.g. how does an evolutionary development approach differ from a spiral model—with concise explanations and helpful diagrams. He also introduces a number of specialized approaches, such as Microsoft Solution Framework, which are widely used in industry but are often ignored in general writings because of their vendor specificity. In doing so, he builds a helpful bridge from academic conceptions of software engineering to the world of software engineering practice.

One of the book's greatest strengths is its crisis-driven perspective. Rather than viewing a "crisis" as a unique event that cannot be anticipated in advance, Zykov treats crises as more-or-less inevitable. And, as a result of this inevitability, he looks at different software engineering approaches in terms of their suitability for handling different types of crises. We all recognize that the large-scale mission-critical system may need to be engineered differently from other types of systems. The book looks at these systems and considers what approaches may be most effective. It also looks at how pure software engineering approaches may be rendered more effective by incorporating techniques drawn from other approaches, such as incorporating incremental prototyping into projects whose overall structure is driven by a waterfall model.

A very important element of the book is the rich systems engineering examples that are introduced throughout the book, and particularly in the later chapters. While there is no shortage of examples in the software engineering literature, far too many of them derive from USA and Western European sources. Many readers, like myself, will be intrigued by the Russian examples and perspectives presented in the book. The similarities and differences between these and the examples familiar to us can help broaden our thinking. As such, the book is well suited to the global audience.

My personal favorite part of the book was the chapter that preceded the final conclusions. It deals software engineering from a knowledge transfer perspective. Because my own area of research is the transdisciplinary field of informing science, I tend to see the informing (i.e. knowledge transfer) aspects of every situation as being central to understanding that context. Unfortunately, much that has been written about software engineering imagines such transfer as being something that will occur automatically (provided the right process is in place). In direct contrast to this, Zykov looks at knowledge transfer processes very explicitly. He even goes so far as to provide a case study involving Innopolis, a Russian city, being built from the ground up to serve as a community that integrates education, research, and practice. Specifically, he considers the challenges that emerged as researchers and educators from Carnegie Mellon University attempted to work with Russian researchers, including Zykov himself, to develop a software engineering curriculum. The entire project makes for fascinating reading.

Zykov concludes his book with the statement:

> In our view, the root cause of the software development crisis is the human mind itself, and we can manage the crisis if we approach human-related and technology-related issues and challenges in a systematic and disciplined way.

Upon reading the book, I felt far more knowledgeable about the challenges facing the managers of software development processes. I also felt more confident in my own ability to discuss the topic (in an informed way) with my students, and in my own writing. It is an important book, one that is well worth reading.

T. Grandon Gill, DBA
Professor and DBA Academic Director
Information Systems and Decision Sciences Department
Muma College of Business
University of South Florida
Tampa, FL, USA

Contents

Acronyms

3D	3 dimensions
6D	6 dimensions
ACDM	Architecture-centric design method
API	Application programming interface
ATAM	Architectural trade-off analysis method
CAD	Computer-aided design
CAE	Computer-aided engineering
CAM	Computer-aided manufacturing
CASE	Computer-aided software engineering
CMM(I)	Capability Maturity Model (Integration)
CMU	Carnegie Mellon University
CMS	Content management system
COTS	Commercial off the shelf
CRM	Customer relationship management
DBMS	Database management system
ERP	Enterprise resource planning
GDP	Gross domestic product
GUI	Graphical user interface
HR	Human resources
IBM	International Business Machines (Corporation)
IEC	International Electrotechnical Commission
ISO	International Standards Organization
IT	Information technology
KLOC	Kilo lines of code
LC	Inductor, represented by the letter L, and a capacitor, represented by the letter C
MSF	Microsoft Solutions Framework
MSIT-SE	Master of science in information technology—software engineering
NATO	North Atlantic Treaty Organization
NPP	Nuclear power plant

PLM	Product lifecycle management
PSP	Personal software process
ROI	Return-on-investment
RUP	Rational unified process
SCADA	Supervisory control and data acquisition
SE	Software engineering
SEI	Software engineering institute
TEP	Teacher education program

Abstract

The book discusses lifecycle optimization of software projects for crisis management by means of software engineering methods and tools. The outcomes are based on lessons learned from the software engineering crisis which started in the 1960s. At that time, software product lifecycles, which the industry had just started moving toward, were anarchic in many ways, since no systematic approach existed. However, as yet, there is no single answer to whether this crisis is over. We analyze the findings of Marx, a pioneering researcher of economic crises. In his terms, a typical reason for a crisis is the "anarchy of production," which results from the absence of central planning or regulation of the production, and of its lifecycle. We conclude that the software crisis might result from inadequate planning, and a lack of common understanding of the product's size and scope between the software producers and the software consumers. The root cause of the crisis is the resource misbalance due to an inadequate, inappropriate, or distorted common vision of the product features and project constraints between the client and the developer. Clearly, each side has a number of roles with very different preferences, goals, and expectations. Crisis in software engineering depends not only on technology-related but also on human-related factors; we take a systematic approach to overcome it. We propose an adaptive methodology for software product development, which optimizes the software product lifecycle in order to avoid "local" crises of software production. A typical methodology includes models, methods, and tools; we start from high-level models for software development lifecycles, which sequentially elaborate the software product. We discuss the general lifecycle pattern and its stages and analyze their impact on the time and budget of the software product development. The model selection influences a number of critical parameters of the software development projects, and it often determines their overall success. Software engineering uses product quality metrics, which make project management more accurate and predictable in a crisis. We identify key advantages and disadvantages for each model; we conclude that there is no "silver bullet," or universal model, which suits all software products equally well. The scope and size of the project are determinants for the lifecycle model selection. We also discuss software development methodologies. These are adaptive process frameworks adjustable to software product size and scope. The methodologies include a set of methods, principles and techniques, and software development tools. Each of the methodologies is flexible enough to implement any lifecycle model. Some methodologies are more formal, others more agile. In crisis conditions, agile methodologies, which usually have fewer artifacts, are applicable. However, agile methodologies require extremely well-disciplined development and impose extra human factor-related constraints. Similarly to the lifecycle models, there is no "silver bullet" for the software development methodologies. However, due to rigorous processes and well-defined deliverables, formal methodologies are better for mission-critical and large-scale applications; in case of undisciplined development, agile methodologies may result in a low-quality product. We move on to patterns and practices of resource efficient software production; this is mission-critical in a crisis. To make product development crisis-agile, we suggest a resource efficient component approach based on high-level architecture patterns for frequently used combinations of large-scale product modules, supported by domain-specific languages and visual computer-aided tools. We approach software architecture in terms of process, data, and system perspectives. We also propose an incremental methodology for crisis-agile development of large-scale, distributed heterogeneous applications. These methodologies provide an industrial level of software quality; they reduce project terms and costs. Implementation areas include oil-and-gas production, air transportation, retail network, and nuclear power generation. We suggest an enhanced architectural pattern for systems-of-systems, which includes five application levels and two data levels. We instantiate this high-level design pattern by functional outlines for systems-of-systems in oil-and-gas and nuclear power industries. We develop crisis-agile models, patterns, and practices for software development and knowledge transfer. To facilitate the transfer, we use informing science and learning principles. Our case studies include knowledge transfer of the masters program in software engineering from the Carnegie Mellon University to the Russian Innopolis University. The key factors, which may inhibit technology transfer, include cross-cultural differences, geographical differences, and process maturity. The key factors which promote crisis transfer include multiple contexts, scaffolding and learning by doing. We recommend reducing the cognitive teaching-and-learning load, establishing and maintaining efficient feedback, building up personal crisis-agile "soft" skills, and addressing higher levels of Bloom's taxonomy. For knowledge transfer, we use an informing framework model, which we enhance by an amplified circuit with controlled resonance and a bidirectional feedback loop. In crisis, knowledge transfer requires special training of the receiving side based on human-related factors. A bidirectional feedback-driven meta-cognitive cycle is critically important for learning quality. We recommend multi-context transfer, which includes hands-on real-world project practice. Thus, we use a multi-faceted approach to software engineering and knowledge transfer, which includes human and technological factors. The systematic approach we use includes formal models, a set of domain-specific methods and visual tools; it increases crisis agility so that software development becomes more predictable, accurate, and adaptive at the same time. In our view, the source of the crisis is a conceptual misunderstanding between the client and the developer, and we can manage the crisis if we address it in a systematic and disciplined way.

Keywords Software engineering · Software crisis · Software development life-cycle · Lifecycle model · Software development methodology · Formal method-ology · Agile methodology · Domain-specific language · Architecture pattern · Heterogeneous application · Pattern-based development · Knowledge transfer · Informing science · "Soft" skills · Systems-of-systems

Introduction. Crises, Crises Everywhere...

The focus of this book is software-related project lifecycle optimization and crisis management; these are based on software engineering methods and tools.

We are going to focus on the reasons for the crisis in software development, which occurred relatively long ago, in the 1960s. We analyze the reasons for the crisis. Software product lifecycles, which the industry had just started moving toward, were anarchic in many ways, since no systematic approach existed.

At that time, software development did not allow precise variation of such basic project parameters as time, cost, or functionality. In fact, the software products were unique masterpieces, with a build-and-fix approach as the core "methodology." Thus, a systematic approach to product lifecycle and responsibility for the deliverables was required. During the following decade, the software development process gradually became a science rather than an art; however, because of imperfect technologies, it had not become a production yet. Large-scale software research and development centers appeared, such as the Software Engineering Institute of Carnegie Mellon University. The value of software increased compared to hardware. Mission-critical software systems appeared, such as military and life-support applications. However, the software crisis lasted much longer and had a deeper nature than that in material manufacturing industries. Currently, the lack of a universal methodology for software development explicitly indicates that the crisis is not completely over. To manage the crisis, we need to optimize the software product lifecycle. Software engineering methods can help in dealing with the crisis, since this discipline systematically approaches software development issues.

The software engineering approach is chiefly oriented on "serial production" of large-scale, complex, high-quality, architecturally heterogeneous, and interoperable software systems. Other architectural aspects include portal-based software systems, remote services, and the like. Software product quality is measurable by a number of "dimensions" or attributes, such as performance, reliability, security, fault tolerance, usability, and maintainability. Heterogeneous software systems incorporate versatile architectures, databases, and data warehouses, which include both strong- and weak-structured data.

The global economic crises and the subsequent depressions taught us certain lessons. We present an adaptive methodology of software system development which allows avoiding local crises, specifically large-scale ones. The methodology is based on extracting high-level common enterprise software patterns and applying them to a series of heterogeneous implementations. The approach includes a new model, which extends the conventional spiral lifecycle by formal models for data representation and management, and by domain-specific language-based CASE tools. The methodology application areas include oil-and-gas resource planning, air traffic control, and nuclear power production. Further, we discuss possible combinations of software lifecycle models and examine the factors for crisis-based terms and cost reduction. Another area of interest is the adjustment of the software lifecycle according to project size and scope. Therewith, the so-called human factor errors resulting from crisis conditions and non-systematic software lifecycles, and their impact on a crisis, are analyzed. The book outlines the ways to systematic and efficient software-related project lifecycles and suggests certain troubleshooting methods.

Another area of concern is the impact of the crisis upon the knowledge transfer processes in software-related projects.

Karl Marx (a prominent German economist of the nineteenth century, Fig. 1) explained the crises and their nature. He stated that the crises result from misbalanced production and the realization of a surplus value on the market [22]. The root cause of this misbalance is the separation between the producers and the means of production.

In terms of Marx, the reason for a crisis is the "anarchy of production," i.e. the absence of central planning or regulation of the production, and of its lifecycle. Decision making, which happens at the enterprise, i.e. producers' level, is often a source of the crisis.

To accelerate capital accumulation, producers over-estimate market demand and push production beyond the threshold of market consumption. Unsold goods accumulate, and the production decreases. For complex products, this often has an avalanche effect on the further suppliers and providers.

Thus, over-investment and over-expansion of productive capacity often triggers the crisis on an enterprise and industry scale. This disproportionate investment and growth between the various sectors of the economy is also a likely reason for the crises of a more global nature. Investments to expand production do not objectively consider the needs of the other sectors of the economy; therefore, over-investment and over-expansion often happen in key sectors of the economy. These key industries include not only material production but also such recent and rapidly expanding industries as software development and IT in general.

Thus, the concerns and expectations of producers and consumers often differ, and the resulting misbalance between the over-production and under-consumption usually results in a crisis. Better planning could help to manage this critical misbalance; however, to plan better we need to establish better communication between the producer and the consumer. This better communication is possible in the case of the dynamical adjustment of the product aims and scope, and only after a clear

Fig. 1 The monument to Marx, Chemnitz, Germany (photograph by the author)

transmission of the product vision to the end user (Fig. 2). The barriers for this common vision are differently focused priorities and expectations. Simply speaking, the customer and the developer use different languages. For certain categories of customers, this difference is even more critical in the case of software products. Thus, the root cause of crises is often a conceptual misunderstanding between the client and the developer. This misunderstanding is often human factor-based, and it could be adjusted with prompt and directed feedback. To overcome this conceptual misunderstanding, we propose a methodology, which includes a set of models, methods, and tools (Fig. 1 in Chap. 5).

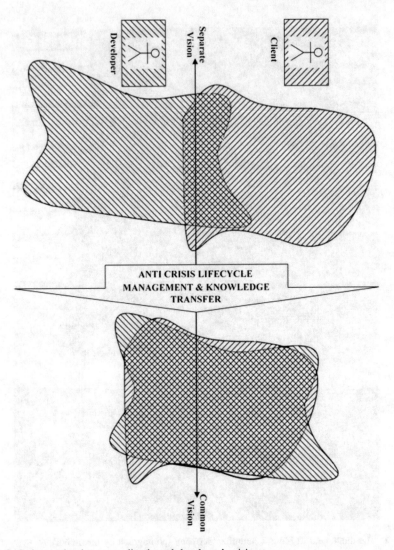

Fig. 2 Software development: client's and developer's visions

We later illustrate this methodology by a set of patterns, practices, and implementations.

This book is organized as follows. Chapter 1 outlines the history of software engineering; it describes software product lifecycles in terms of models, methodologies, and tools. Chapter 2 discusses the strategies of software lifecycle optimization. Chapter 3 compares a number of adaptive software process frameworks and their applicability to crisis management. Chapter 4 addresses crisis-agile pattern-based software development. Chapter 5 focuses on the human factors, which promote software-related knowledge transfer in crisis. The conclusion summarizes the key findings of the book; it suggests possible ways to manage software development crises.

Chapter 1
Software Engineering: The End of the Crisis?

Abstract The chapter gives an overview of the concept of crisis and its application to software development and software product lifecycle. Managing software development lifecycle is a challenge for large-scale and mission-critical applications, especially in crisis. To solve the challenge, a lifecycle management methodology is required; the methodology includes models, methods and supporting tools. We discuss the general lifecycle pattern and its stages. We find that the cost of defect detection and fixing increases exponentially as we move from the earlier stages to the later ones, so error detection should happen as early as possible. Lifecycle models selection determines the mission-critical parameters of the project: the architecture of the project, its budget and timeframe. The model selection also determines product artifacts and quality attributes; these are based on product quality metrics, which make crisis management more accurate and predictable. The lifecycle model selection should be adequate to the experience of the project team in terms of problem domain expertise and operational knowledge of technologies, tools and standards. We describe a number of lifecycle models, such as build-and-fix, waterfall, incremental, object-oriented and spiral. Some of these models require iterative development; the others are more straightforward. Certain models require a high level of discipline and organizational maturity. There is no universal model, which suits any software product equally well. The scope and size of the project are the determinants for lifecycle model selection; we can customize or combine certain models in order to adjust to the specific features of the project. For each model discussed, we identify the key advantages and disadvantages.

Keywords Software engineering · Software crisis · Software development lifecycle · Lifecycle model

1 Introduction

As we understand from the introduction to this book, crisis management of software development is a challenge. In this chapter, we are going to look in more detail at the concept of crisis and its application to software development.

© Springer International Publishing Switzerland 2016
S.V. Zykov, *Crisis Management for Software Development and Knowledge Transfer*, Smart Innovation, Systems and Technologies 61, DOI 10.1007/978-3-319-42966-3_1

To cope with the crisis, we need a methodology to manage the software development lifecycle. Such a methodology should wisely combine models, methods and tools.

We will present a brief review of a number of lifecycle models used in software development. We will discuss a general lifecycle pattern and its stages, from product concept to retirement.

As we will see, selection of the lifecycle model influences a number of global parameters of the software development project, such as budget and time, and it often determines its overall success and the resulting product quality. To make crisis management more accurate and predictable, we can monitor and manage the above global parameters by means of certain product quality metrics.

We will compare each lifecycle model in terms of strong and weak points, technological and business constraints, and required levels of software development discipline and organizational maturity.

Clearly, there is no universal model, which equally suits any software product. However, a wise combination of the lifecycle models and their adjustment to the project scope and size is usually one of the keys to success in software development, even in crisis conditions.

This chapter is organized as follows. Section 1 discusses the similarities and differences between software development and material production. Section 2 presents a brief sketch of software engineering evolution. Section 3 describes the engineering of software lifecycle. Section 4 provides an overview of the software lifecycle models, development methodologies and supporting tools. The conclusion summarizes the results of the chapter.

2   Software Development and Material Production: The Same or Different?

Let us focus on the reasons for the crisis in software development.

The crisis in software development started relatively long ago, approximately in the 1960s. Let us analyze the reasons for the crisis.

At that time, software product lifecycle, which the industry had just started moving toward, was anarchic in many ways, since there was no uniform, systematic approach to software development in terms of product and process. Software development did not allow precise monitoring and adjustment of the major project parameters, such as time, budget and product quality. In fact, it was hard to manage the software development, as the build-and-fix approach was the key methodology. Because of the absence of proper standards, metrics and rocketing complexity of the computer hardware and the related software development tasks, the software development lifecycle was hard to manage, predict and control.

The following decade of the 1970s showed that the software development process was gradually becoming a science rather than an art; however, because of imperfect technologies it did not become a production process yet. The era of hand-made software products from unique gifted programmers was over.

A few large software research and development centers appeared; one of the most well-known examples was the Software Engineering Institute of the Carnegie Mellon University [1].

The value of software, as compared to hardware, increased tremendously. Mission-critical software systems appeared to monitor and control military and life-support applications.

However, the software crisis, which started in the 1960s, lasted much longer and probably had a deeper nature than that in material manufacturing industries, such as construction, automobile production and the like. The absence of a relatively universal methodology, the so-called "silver bullet" for software development, explicitly indicated that the crisis was still there.

To manage the crisis, we need to optimize the software development lifecycle by systematically approaching all of its processes and the resulting product quality attributes. Since software engineering addresses the software development issues in a systematic way, we assume that its methods and tools are probably useful for the crisis management.

To understand the nature of crisis, we will analyze a number of patterns for software development lifecycle, each of which has certain phases or stages. Of these, we will identify the most mission-critical phases in terms of resource consumption and management dependencies, and see in what way they are different from the material production. We will see that irrespective of the product size and scope, software production has a number of fundamental differences as compared to material production. For example, software development lifecycle is often essentially shorter than that of a material object, since software products usually morally degrade much faster than the material ones. Another consideration is that the software product changes are often more serious and radical than the changes in material objects. For instance, there is a number of buildings and bridges used for many decades at relatively small or even negligible maintenance costs as compared to software.

Brute force approach is not quite applicable to software. For example, doubling a data channel throughput would not guarantee its reliability. However, making a bridge trestle two times thicker would result in a deliberately reliable product.

In contrast to material production, a software product model and its prototype should not necessarily be reliable. Any operational software release usually contains a certain number of defects; however, it is not catastrophic for the product, as it is in the case of a certain automobile model. Even after a serious software crash, it is often sufficient to just restart the product rather than to repair or reproduce it. Software defects tend to accumulate in time. Defect detection is a challenge, and building defect-free software systems requires different methods as compared to material objects construction.

In addition to the lifecycle patterns, or models, we will consider the software development methodologies, which include processes, roles, methods and tools. According to each process step, which consists of larger workflows and smaller activities, every role is to produce a certain artifact or deliverable that is a part of the software product. These deliverables include not only software code, but also documentation artifacts, which are equally important for the product quality.

Software product development usually requires a team, especially in case of large-scale and complex products. However, the accumulated team experience in software development does not always result in system quality increase. This often happens because of rapid changes of complex software platforms; another important reason may be the human-related factors, which depend on the so-called "soft" skills, such as communication, negotiation, self-reflection and self-adjustment.

The software development lifecycles are often iterative, incremental and reuse-oriented. However, we will also consider straightforward one-pass and abbreviated processes, which proved crisis efficiency under certain conditions.

Concerning the software engineering methodologies, we are going to analyze a number of process frameworks for software development. Though all of them are potentially adjustable, we will identify and compare formal and agile methodologies in terms of crisis applicability for general-purpose and mission-critical software product development.

We will then discuss patterns and practices of the methodologies application to real-world implementations. These embrace several industries including civil air transportation, oil-and-gas production and nuclear power plant construction, to name a few.

To address the human-related factors, which may distort common vision of the product by the client's and the developer's sides and thus initiate a crisis, we propose a number of models and techniques. These include informing science approach for efficient communication modeling, and a set of psychologically approved practices for knowledge transfer management.

Software development and material production have a number of similarities and essential differences. Thus, we need to use software engineering models, methods and tools to provide predictable and measurable product development in terms of quality. This is mission-critical for industrial production of large and complex software systems with such quality attributes as availability, maintainability, reliability, security and reusability. In crisis, however, software engineering methods and tools provide a rigorous technology basis required even for general-purpose software production.

3   Software Engineering: Start and Evolution

The focus of software engineering is serial production of high quality, relatively complex and usually large-scale software systems [2, 7]. Such software products often manage terabytes of data, thousands of files, and hundreds of components; they are distributed and usually support multiple users, virtualization, and large databases.

In software engineering, the quality of software products is usually measurable in terms of several dozens of quality attributes. The ISO/IEC standard on systems and software engineering describes system and software quality in terms of reliability, security, fault tolerance, usability, reusability, maintainability, and a number of other measurable attributes [3]. We may use these attributes for optimizing the software development lifecycle, which is based on software engineering models [4–12], methods [13–18] and tools [19,20]. The idea is to identify and prioritize critical quality attributes for crisis efficient software product development.

Further, based on the constraints delimited by the quality attribute values, we may follow a disciplined process of sequential elaboration of the software product functions, which includes conceptual modeling, analysis and design, prototyping, implementation and maintenance.

Let us treat constructing a software system as an industry-level production process for a large and complex material object, such as an automobile, a bridge, or a skyscraper. Although there is a certain similarity between material and software product development, a number of significant differences also take place.

In the late 1960s, the participants of the NATO Software Engineering Conference drew the following conclusion [21]. The conference showed that software should be built according to different principles than that for material products, since their lifecycles are fundamentally different in a number of ways. However, the major software project and product parameters, such as project terms, product cost, size and quality can be measured formally, and software development lifecycle can be managed by means of technologically approved methodologies.

Let us overview the start-up of the software engineering as a discipline.

A number of factors influenced the foundation of software engineering as a discipline. One of the key events, which triggered the software engineering advent, was the historical NATO conference, where the relation between software engineering and material production had been clarified [7]. The major conference decision was that regardless of the fact that software engineering had much in common with material production, this new research area had a number of clearly distinct features, which required a different approach to lifecycle and development. Another important decision made immediately after the conference was that there was a specific kind of crisis in software engineering.

Software crisis is a term used in the 1970s to indicate the challenge of producing useful and efficient computer programs in the required time. The software crisis happened because of the rapid increase in computer power and the complexity of the problems. The software complexity increased as well, and a number of development problems arose because the methods existing were low quality ones.

The term "software crisis" was coined by F. Bauer and some other participants of the first NATO Software Engineering Conference in 1968 at Garmisch-Partenkirchen, Germany [22–24]. Later on, F. Bauer recalled that he "… discussed the so-called 'software crisis'" with E.J. Helms, the Danish representative of the same NATO Science Committee.

In the 1970s, the term became so influential that E. Dijkstra, the Dutch representative of the NATO Science Committee, referred to it in his Turing Award

Lecture: "The major cause of the software crisis is that the machines have become several orders of magnitude more powerful!" [25].

Thus, the crisis resulted from the rocketing complexity of the software development processes. The immediate crisis effects were project management challenges in terms of budget and time, and software product inefficiency in terms of quality and maintainability. Not only did the software products seldom meet the clients' requirements, but also they often were undelivered at all. Increase of the computing power outperformed the developers' abilities to use the emerging resources efficiently; one of the examples of such a computing power boost was the revolutionary IBM B-5000 computer [26]. A number of models and methodologies appeared to improve software product development; we will discuss them later. However, development of large-scale software products under rapid requirement changes, poorly managed processes and budget uncertainties often caused complex and "wicked", i.e. unanticipated problems.

The same F. Bauer, who coined the "software crisis" term, also suggested the "software engineering" approach as the remedy. After the historical 1968 NATO conference, he initiated another meeting, where the prominent computer scientists (and the NATO Science Committee members) from US and Europe had a new round of discussions on the crisis and the means to manage it [27].

The birthplace of the software engineering education is Carnegie Mellon University (CMU). The Software Engineering Institute (SEI), which is interwoven into the CMU educational processes and activities, was founded as a result of the 1968 NATO conference decisions. Historically, the SEI focus was research and development of large-scale software systems with heavy duty and high reliability. The primary customer of these software systems was the US Department of Defense. Since SEI was the developer of standards in software engineering, such as SWEBOK [28], the CMU educational standards were probably most close to the industry. However, CMU had a long way prior to SEI came into being, and the initial CMU approach essentially influenced the techniques of training in software engineering.

The CMU was founded in 1900 as a synthesis of Carnegie School and Mellon Institute, which were among the leaders in US research in technology and economics respectively [29]. The synergy of this new blend was so powerful that the venture became a leading university in the area shortly.

In the 1800s, Andrew Carnegie, who came from Scotland to the steel producing area, founded the Carnegie Technology School. Among his key ideas was the learning-by-doing approach. This hands-on starting point helped him to train the steelers' children in a college-like environment. The idea was to deliver the just right amount of knowledge to master the innovative engineering technologies. The approach was practically oriented and the deliverables were real-world engineering systems and projects. The main Carnegie's idea still holds true; it leads to realistic, well-justified solutions for heavy duty software systems with a solid engineering-based reasoning. Such justification is based on rigorous software engineering metrics, which guarantee development and maintenance of "good enough" software systems in terms of availability, performance, modifiability, security, usability and a number of other quality attributes.

One more ingredient of the CMU success was an early adoption of the cognitive approach to software engineering, which has been chosen due to tight integration with a number of psychologists who contributed to the foundation of the famous CMU School of Computer Science [29].

The general framework of CMU curricula is based on the above-mentioned factors of learning-by-doing, Carnegie and Mellon schools alliance, and interaction with SEI and School of Computer Science. Further, we will discuss masters' software engineering and IT courses, which are among the primary concerns of the knowledge transfer for the recent Russian ambitious Innopolis University project.

During visiting faculty training at CMU, we examined the following core courses of the master in IT and software engineering (MSIT-SE) program: Analysis of Software Artifacts, Architectures for Software Systems, and Personal Software Process (PSP).

Chapter 5 summarizes takeaways from the above-mentioned courses and suggests certain ideas on their application to the Innopolis University curricula.

4 Software Engineering and Product Lifecycle

Software engineering is an area of computer science that deals with the construction of software systems, i.e. a number of interacting software components that are so large or complex that construction of such systems requires participation of a dedicated development team or several interacting development teams [30]. Under the developers we imply not only programmers or coders, but also system analysts, project managers, testers, system architects, documenters, quality assurance people and maintenance personnel. That is quite a big team, which focuses on the production of a software product in an existing environment of software systems of the customer. Software engineering is therefore vital in terms of management of all levels and all aspects of software development, such as requirements analysis and specification, preliminary and detailed design, implementation and testing, integration, transfer to client and customer support.

In *Software Engineering*, Lipaev, the patriarch of the Russian software engineering, gives the following definition: "The software engineering refers to a set of tasks, methods, tools and technologies, which is intended for design and implementation of complex, scalable, replicable, high-quality software systems, possibly including a database" [31]. Each word in this definition is specifically meaningful for crisis conditions and mission-critical systems.

According to the above definition, software engineering as a branch of science is precisely aimed at creating mission-critical software systems. Because these software systems are complex, it is not economically feasible to immediately replace all their components that are used by the large-scale software subsystems. More likely, particular subsystems of a large-scale system are reused, since they are high quality by design and often replicable. Examples of such solutions include Microsoft Dynamics and Oracle e-Business Suite. Under scalability, we mean a gradual decrease in performance in case of intensively growing system load. In addition, these systems should be reliable, predictable, ergonomic and maintainable, i.e. they should be developed to provide a sufficiently flexible and relatively evolutionary interaction with the user at the stage of pilot and commercial operation. The above definition of software engineering includes "design and implementation", i.e. it follows the core of the software development lifecycle. An important addition is that a large-scale software system often contains a database. These databases can be heterogeneous, i.e. they can include object components, so they are not purely relational. For example, recent versions of Oracle DBMS are referred to as object-relational. Other "new generation" databases, such as O2, Orion etc., combine relational and object-oriented paradigms.

Software systems development generally includes such concepts as software project and software product. Currently, we are going to focus on the software product, i.e. to look at the lifecycle from the perspective of software system architect. Software project is the perspective of the project manager, who is responsible for managing the project team, communication of the people involved in the project, time and cost. On the other hand, concerning software engineering we can generally discuss product development for a specific customer. However, we should plan all processes and technologies related to the software development in such a way that it is possible to supply the product to a wider audience of consumers, and ideally to make it reproducible and commercial off-the-shelf (COTS). It is advisable to provide a high percentage of reusable elements of the product, such as code, documentation, database structure and software architecture, so that after final release is ready any possible customer could spent minimum time, budget and labor to customize it. This approach is essential for crisis management of software production.

At the initial stage of product development, as a rule, there is as little as only a concept or an idea. Of course, at least a minimum amount of initial investment is required. However, in case of project development, a draft, high-level plan should exist, which delimits such key indices as budget, functionality and time, and there should exist a certain customer, i.e. a specific shareholder who will provide funding for the project. In some cases, we can build software by mixed development, so that a proprietary system can later be transformed into a relatively open solution, which suits a wider range of customers.

What are the main features of a software product? First, it should have a certain commercial value. This means that the product is intended to solve a specific problem for a particular class of end users, clients or consumers. Thus, the product should be supplied to the market in order to meet specific needs and custom business objectives. What are the examples of such software products? Often these are physical objects, such as an information media, e.g. DVD, CD etc. However, these might also be non-material objects. In any case, a software product should include proper customer documentation and a number of legal agreements, such as a license, a partnership agreement, and so forth. A software product can also be offered as a service for deployment, customization, maintenance, or consulting.

Software products can be classified on different bases. One type of classification is the scale or scope: it is for personal use, non-commercial, or comes as a commercial COTS product for a wide range of organizations or individuals. Another method of classification is the purpose of end user. In this case, we can divide products into specialized software aimed at solving relatively specific tasks, such as software developed to solve astronomical problems, laser ranging, and more general-purpose products, such as the operating system, office software etc. One more type of classification is the degree to which the product is open for interfacing with others. In this respect, the software can be classified as ready-made proprietary products and customizable component-based products, such as an API library.

Any software development takes place according to a certain lifecycle pattern, which includes a sequence of stages; it generally begins with basic concepts and ideas, and it generally ends with the product retirement.

The concept of lifecycle is applicable to any kind of systems, for example, such systems as skyscraper buildings; however, the lifecycle of software products has its own distinct characteristics. Software development is usually a gradual evolution, which starts from the initial concept or a rather abstract idea. It is further elaborated into a piece of software, which includes not only code but also a large number of documentary artifacts, such as inline code documentation, specific documentation for software administrators who setup, install and maintain the product, and others. The lifecycle of a software product ends at the stage of retirement, which follows the maintenance.

Each stage of the lifecycle is completed after developing a certain artifact for the system. Depending on the model lifecycle, after each lifecycle loop, the system can be either fully functional or not full-featured. Each stage ends with the production of documentation, which may include global artifacts, such as a project plan, a test plan, an implementation plan, a maintenance plan and more specific documents, such as use cases, Administrator's Guide, Quick Start Guide, high-level requirements of the product, or more detailed requirements in the form of technical specifications. Size, nature and elaboration of the documentation artifacts depend on the scale and scope of the software. Of course, each stage of the production of software should be clearly defined by its start and end time points, as well as by deliverables for the next stage in terms of code and documentation. In practice, however, mission-critical software product development is often more complicated; however, the software engineering approach, even in a crisis, requires a thoroughly defined lifecycle, each stage of which should yield to new product artifacts including new documentation.

To study the lifecycle of mission-critical software systems, we should first understand how software development is organized, i.e. how the critical software development processes are associated with the lifecycle stages. Failing to understand the lifecycle in general, we can hardly speak of any systematic organization and management of these processes. Of course, the most successful projects have to draw conclusions and replicate the principles that led us to success. We should study and improve the practices and techniques that allow for efficient and systematic development of mission-critical software in order to improve the product operational quality, user interface, documentation and all the related processes that underlie the lifecycle. We should do these improvements based on the analysis of the historical data for the previous projects. These should guide us in the future project planning and creating other "global" documents, such as testing plan, integration plan, implementation plan, maintenance plan etc. We should also use the error reports, and other documents created during software product testing in order to improve the processes and make them manageable even in a crisis. In this respect, thorough study of product development lifecycle provides an important basis for forming anti-crisis patterns of software development, which allows more accurate planning, monitoring and reproducing the processes of the lifecycle. Therewith, we need a methodology, which is designed for scalable teams of developers, and which makes it possible to adjust and adapt the lifecycle in a crisis, and to achieve an adequate product quality. Each software development project should follow a certain procedure, i.e. a methodology, which includes all the previous experience and historical data available, and which must adapt to the nature, size and scope of the particular customer and the specific conditions of production, including specific crisis conditions. Oftentimes, the customer already has a certain and a unique combination of the hardware and software environment in which the new software product must be implemented. This is particularly important in relation to large-scale and mission-critical systems, due to a large number of relationships and significant complexity of the software environment. Thus, thorough analysis and planning of the lifecycle is a requirement for any mission-critical software product, and it is essential in case of crisis.

When talking about the lifecycle, we need to make some important remarks. First, we have to say that the lifecycle processes, which embrace each stage of the software product development, include a number of parties. At a minimum, these are the customer's representatives and representatives of the developer and management. The representatives of the customer, who are going to accept the product, are often technically competent people. As a result, they are doing the quality assurance. The developer's side includes a wide range of experts, such as analysts, risk managers, designers, system architects, documenters, programmers, testers, maintenance specialists and more. The management side often includes project manager, product manager and others. It is clear that the client's and the developer's employees have very different business goals. Similarly, the attitudes of the management of the client and developer differ in many respects.

These different perspectives are due to differences in product expectations in terms of functionality, design constraints, deadlines, cost, functionality, and they often result from different interpretations of certain terms and conditions. The customers expect software product to be good enough for assisting in their business needs; however, they often can be unaware of sophisticated technologies that support the features of the software product, which are clearly understood by the developers. In this respect, even reasonable views, approaches and attitudes towards lifecycle requirements and restrictions may appear to be very different for the customer and the developer sides (including their various representatives at a number of levels), which can lead to negotiation problems and to a significant increase of the project time and cost. Coordination of these problems between the developer and the customer becomes a matter of great importance: it allows to arrive to a common understanding of the key design constraints, and in many cases helps to avoid the crisis, which is often human factor-related. Therewith, the customer usually wants to impose the bottom limits on the key product features, for example, requesting that the number of concurrent users must exceed a certain value. Thus, the software developer should be able to arrive to a written agreement with the client on these mission-critical parameters. This agreement may be a legal document contract, such as technical requirements, requirements checklist or some other document. The software product developer, in contrast to the client, often seeks to impose the constraints on the top limits of the product operating parameters, such as number of concurrent queries and throughput, or to verify that the product is still going to behave adequately under the technological and financial constraints, so that it will operate within the required performance range.

In addition to the above listed categories of the developer's side, we can mention the following essential roles: portfolio manager, team leader, analyst, chief architect, subsystem designer, usability expert, tester, coder, test manager, and technical writer. These roles represent only the key classes of project team members, and these classes are often instantiated by a large number of participants in mission-critical projects. The interaction between these roles is a challenging task in terms of project and product management. Further, we are going to focus on product management.

In case of crisis, what is the major aim of the developer of a mission-critical software product? In short, the main objective of the developer is to create a "good enough" product. However, what does this "good enough" mean? We are going to discuss it together with the mission-critical factors of crisis software development.

It appears that software product development is a multi-factor optimization because, in fact, developers need to negotiate with the customer on the outlook and set of requirements for the product. This will be the major starting point, which will be elaborated in order to become a software product with a proper documentation. Given such an initial point as an input of the software development product, we can arrive to quite a number of possible outputs, because there is an infinite number of possible variations of code and documentation that meet the requirements of the customer. However, a realistic software lifecycle should be adequate and predictable in terms of deadlines, budget and functionality, and it should provide a choice of lifecycle options. It is required to identify the phases, their boundaries and an adequate number of the iterations in order to get the required functionality and quality of the software product. In other words, the process of software development is a multidimensional, multi-factor optimization, which usually takes into account at least the following criteria: the terms of the project, the cost of the product, and the quality of the product, which includes both documentation and code. The quality attributes of the software documentation include traceability for entire documents, separate artifacts and items, so that they are valid and meet each other in terms of conformance, completeness, consistency, integrity and compliance with the original problem specification. Maintainability is another possible critical optimization factor, since it provides essential cost reduction of the most resource intensive part of the product lifecycle. The priority of optimization factors is not pre-determined, however, it is dependent to a large extend upon the nature and scale of the software project. Let us assume that the development time for a small-scale system is usually below 10 person-years, for a medium system it is normally between 10 and 100, and for a large one it often exceeds 100 person-years. For example, enterprise systems development time usually amounts to 100 person-years and above. That means a very high development cost, which is another crisis factor; however, it also means that we should look for opportunities to save even more in terms of labor, cost and time while implementing an enterprise application.

A software product and a software project are clearly different concepts; the stages of the product lifecycle are somewhat broader and include feasibility study and conceptual framework of the project, i.e. the key idea, which it starts from. Lifecycle of a project is largely completed after the transfer of the product release to the client. However, lifecycle of a product usually includes such phases as maintenance and retirement.

Further discussion of the economics of software product lifecycle includes a number of steps, which are important in terms of contribution to profits and sales [32–35]. At the initial stages of construction and introduction to market, the product is usually negative in terms of profit. However, after the market launch, as the sales usually rise and the product matures, the profits often become positive. At this stage, the product generally has a positive profit value though its profits decrease. Later on, product decline happens; it is characterized by a significant dropdown and by relatively low absolute values of the profits and sales. This is typical both for a COTS product with a large number of installations, and for a proprietary product tailored for particular customer.

Let us look closely at the lifecycle economics based on comparison of development criteria, rate of business growth, and market share of the software product. In the beginning, we need investment, and the future of software is uncertain. Then the software product enters the market, generates income and profits, and sales support expenses are insignificant. After a certain period, there comes a stage when profits are relatively low and sales require significant expense.

If we recall the description of the software lifecycle stages, we can see that there is a number of steps virtually independent of the applied methodologies for software systems development. These steps include analysis of requirements for the software product, specifications of a software product design (high-level, preliminary, detailed, and operational), implementation, testing (unit, partial, and full product, acceptance), integration, maintenance and retirement.

Documentation is an important component of any software product. A common misconception is that the documentation is optional or it can be neglected. Documentation is an important deliverable for a software product of any size and scope. Let us assume that a software product developer tries to recall or understand his or her own code a few years after it has been built. Naturally, it is very hard without a proper documentation. At the stage of product operation, the maintenance team reads someone else's code, which is quite challenging. It is so because the person who reads the code usually has an average knowledge of programming, and due to the fact that the maintenance personnel is usually out of the code construction. However, as a rule, this is what happens. It becomes clear that without a proper documentation it is practically impossible to read and to understand any code. Very often, a project team stays together only to create a certain software product, and after the project is complete the team members never meet each other. Thus, code maintenance is successful only if the product comes together with the supporting documentation. The role of documentation quality is very important, its cost usually pays off, and proper documentation provides flexible maintenance. After maintenance stage is complete, there comes product retirement. Documentation often drives the software development lifecycle. For example, the documents produced at the requirements analysis phase become the basis for the project specifications document, which in turn is the basis for the design stage. Design documentation usually includes a number of diagrams, such as use cases, class diagrams and the like, which make the basis for the implementation and the following stages of software product development. Thus, documentation is an inherent part of each stage of the software development lifecycle.

Let us give more details for each stage of the software lifecycle. The first step is requirements analysis. There is a meeting between the representatives of the developer and the customer. The aim is to achieve a common understanding of the problem itself and of the software solution that is required to meet the business objectives of the customer. Of course, in some cases the customer may not have a complete understanding of the technological features of the project including software construction processes, experience of the software development team, state-of-the-art technologies and standards for the product design, implementation and transfer. Often, the customer is not technically literate; however, he or she usually has a clear idea of the particular problem domain of the future software solution. Similarly, the developer often has a limited understanding of the characteristics of the problem domain. For example, in case of oil-and-gas domain, it may be important to present the results of research of seismic activity of the earth crust, including the dynamics of three-dimensional representation of the geological data. Since this is a very specific type of data, which is hard to adequately perceive and analyze by an average development team, the developer's management should probably include a geology expert into the team.

In other areas, such as the coal mining, geological data has other distinct features, which are different from the previous oil-and-gas example. Therewith, it may be difficult to reach a common understanding of the challenges, especially the specific features of the problem domain for which the software product is intended. Thus, a very important objective is to identify and discuss the entire set of the high-level functional and non-functional requirements also known as quality attributes. Moreover, the constraints for software product requirements should preferably be quantified in terms of metrics. To reach the common understanding of the mission-critical requirements, a series of interviews is often required. The result of these interviews and meetings is a document that contains a formal description of software requirements as a list of requirements and technical specifications. This result is fundamentally important, especially for crisis conditions and mission-critical software systems. The subsequent lifecycles steps—design, implementation, integration etc.—follow the requirements document; they include the functional requirements, quality attributes (performance, availability, security, usability and so on) and the quantitative constraints for the software.

The next software development step is the project specifications. It is based on the requirements documentation, i.e. the deliverables from the preceding stage of the lifecycle. This step and the following ones contribute to responsibilities of the developer. Specifically for crisis conditions, there is a number of "flexible" or "agile" methodologies for software systems design and implementation. Examples of these include XP, Scrum, Agile. According to these methodologies, the customer is actively involved in all stages of the software development lifecycle. For large-scale and mission-critical systems, as a rule, software development lifecycle is based on more formal processes such as Rational Unified Process (RUP) or Microsoft Solutions Framework (MSF), where the client's role is more passive, and the main actor is the developer. Design specifications contain a description of the functionality of the entire product and all its major constraints; it is desirable to express these constraints quantitatively. At this point of time, as early as possible in the crisis lifecycle, we should specify the software technologies and architecture, and to make a choice between the platforms, such as Java or .Net. It is critically important to specify explicit values for the number of concurrent users, connections, transactions and their intensity, bandwidth/throughput, and some other parameters. In crisis, the methodology and the lifecycle model of the software development should also be chosen as early as possible, since the choice of methodology and model lifecycle has a critical impact on the timeframe, budget and success of the project. Project specifications must limit the time and cost of the project based on the agreements between the developer and the customer achieved earlier on in the lifecycle.

Further, based on the preliminary design specifications, detailed design is produced, which describes the software architecture in terms of components of the product and their interfaces represented as connectors. In case of object-oriented lifecycle model, which may use significant parallelism for crisis agility, the modules and interfaces between the components should be integrated into the software environment of the customer. However, in large-scale products, there is often a certain amount of interacting systems already in operation at the client's site. That is why the developers should take into account the conditions of the current hardware and software environment of the client. In case of a mission-critical system, the client can use a complex set of servers, such as database servers, cache servers, security servers, telecommunication servers and so on. The developer is also responsible for detailed design phase of the software lifecycle. In addition to architecture diagrams for the major parts of the software product, usually represented in terms of components and connectors, detailed design output provides a high-level document that describes the large-scale software system components and connections including integration points of the new product with the existing client's environment.

After detailed design and product review, i.e. checking the product specifications for the internal correctness, completeness, consistency, integrity and for compliance with the technical requirements, the developer usually can proceed to implementation, i.e. code generation for the software product and the accompanying documentation.

During software coding, any modular product is elaborated as prescribed by the components and connectors identified in the previous lifecycle stage. Developer, based on the detailed design documents, does the product implementation. Developer also takes into account the general project plan, since it is necessary to make important decisions on the limits of testing, schedule of individual modules production, and transition to integration and the subsequent stages. The lifecycle pattern choice determines the success of the transfer to the customer and the quality of the software product. Therefore, the total project plan, which includes global limitations of the time and budget, as well as the most important functional parameters and constraints of the software product, must be taken into account at this stage to ensure the correctness, predictability and quality of the implementation process.

Implementation is also a stage of the developer's responsibility; coders and testers are involved in this process. At this stage, individual modules are developed. These relatively small parts of the software system solve relatively independent tasks. The previous stages have already identified basic parameters for the modules, such as algorithms, data structures, local and global variables. For instance, in case of object-oriented lifecycle model the class structure is identified in terms of main attributes and methods. The result of this stage is a set of individual software product modules, each of which was separately implemented and tested by the developer in order to comply with the internal correctness and design specifications. At the stage of implementation, developers produce certain types of documents related to testing, such as unit test cases, and the other product documentation for each module. The project documentation includes descriptions of the modules, their functions and interfaces, their interaction with other modules, their key features (including attributes, methods, algorithms and data structures) and the documentation to the code, which allows for code review and analysis without direct code execution.

After production of the individual modules, which have been tested for integrity and quality in terms of allowed error threshold, developers can proceed to the next lifecycle stage, which is integration. At the integration stage, developers assemble the product according to the high-level architectural diagram, which was produced in the architectural design phase, and test its fragments. Typically, the modules are tested in pairs or larger sets, thus forming partial products, and finally the product is complete. After that, the developer and the customer conduct final testing, and product transfer takes place based on the software product acceptance tests.

Acceptance testing stage is usually the first time when the software product is installed at the customer's site using the actual software and hardware environment, and the actual data. The data and environment must meet the actual operating conditions of the customer's software systems. In case all acceptance tests are successful (i.e. the product follows functional requirements, it is usable and fits into the environment of the customer) the transfer of product occurs, and the phase of maintenance begins.

In terms of lifecycle economics, maintenance is the most expensive stage, which often takes around 2/3 of the product costs [7]. However, the maintenance stage is a requirement for any software product. This is so because the goal of software product development is continuous and productive relationship with the customer rather than merely software transfer. The objectives of the software maintenance are to fix the defects that remain in the software product, to update the product according to new project specifications, to improve performance and to accommodate to changes in the software and hardware environment of the customer.

Maintenance usually includes the following activities:

- Corrective maintenance, which fixes the existing defects in the software product without changing the design specifications;
- Perfective maintenance, which implements changes to the product functional requirements, making the new product release with improved functionality and same or better quality in terms of performance, reliability, security, availability, usability etc.;
- Adaptive maintenance, which is software system modification in order to adapt the product to the new software and hardware environment.

After completion of maintenance, comes the stage of retirement. This takes place after complete termination of the software product operation. However, if certain product functions are still required by the customer, the data from the previous product should be exported to the new software systems prior to retirement. The cost of replacement includes the cost of technology changes, the new software product development costs, maintenance costs for the new software, the cost of personnel training to use the new software and technology, and the cost of short-term dropdown in performance during the period of technology replacement.

Let us consider maintenance and its artifacts in more detail, since maintenance is most expensive and thus mission-critical lifecycle stage in terms of crisis.

Maintenance starts after successful acceptance testing of the software product. The customer conducts acceptance tests, or at least actively participates. The product should be tested with the real environment of the customer, i.e. software and hardware, and real data in terms of size and content; the results of each test must be successful. After acceptance, the entire software product including the code and the documentation is transferred to the customer. According to object-oriented lifecycle, the product documentation includes class diagrams and use case scenarios, which describe the basic functionality of the product and its behavior in different conditions. The product documentation also describes the main product modules and their functions in terms of methods and interaction between the classes and the environment. The product documentation includes class signatures describing the basic functionality, interactions with adjacent modules, local and global variables, data structures and algorithms. The product documentation also contains instructions for installing and running the software product, diagrams that describe behavior of its architectural components, and a number of other artifacts. The documentation also includes user-friendly manuals for different categories of the end users of the software product. The user documentation often contains a brief description of the basic functionality as well as a complete user guide, which includes description of the errors that typically occur when operating the product, use cases for its functional modules along with the screenshots, and the glossary of frequently used terms.

In order to manage the crisis, we recommend to reduce the maintenance cognitive load by separate application of each kind of maintenance mentioned above.

Corrective maintenance is a required activity for any software product. It includes elimination of residual failures, i.e. significant defects in the implementation of the software, which remain in the product after the acceptance tests. Another source of residual failures is the actual product operation by the customer. Naturally, despite the fact that the number of defects decreases exponentially during testing, it is impossible to eliminate all defects. Thus, end users may identify a significant number of serious defects while working with real data under a complex set of scenarios. The corrective maintenance does not address minor defects, but specifically the defects that result in a halt of the system, a critical failure, or a loss of data, so that it becomes impossible to continue the product operation, and so on. The corrective maintenance does not include changes in the functional requirements; adding new features to the product requires perfective maintenance.

The aim of perfective maintenance is to add new functionality. While operating the software product, the customer often concludes that some of the features were not included into the original design specifications. This can happen for various crisis-related reasons, such as budget problems and development schedule. During the maintenance phase, a functional update of the product is often required. This may result in a new iteration of software development, production of a new release of the current product or even a new product. In terms of the contract, a supplementary agreement is produced, which specifies this new functionality.

Another type of maintenance is responsible for quality improvement. It is useful when the customer is satisfied with the functionality that is already implemented; however new non-functional requirements are required in order to increase performance. For example, in case of an online store, a problem could be the internet throughput because the actual number of users is significantly more than the planned one. The new technological constraints require database server change for a more data intensive one; the new software product should support transactions, and the overall performance of the system in terms of response time should remain at least at the same level as before.

Adaptive maintenance is related to the migration of an existing software product into a new environment. Under the software environment, we understand the entire set of software systems available to customers, such as operating system, database server, and the like.

Let us briefly discuss what is necessary to ensure the enhancement of the software products with the features provided by the software or hardware environment.

We give a more detailed discussion of that in the section describing the enhanced spiral model in Chap. 2; we also present the enterprise crisis agility matrix in Chap. 5. The general idea is to develop software in such a way that it is maintainable. Maintenance is a required stage of lifecycle of any software product, no matter how small it is and whatever is the developer's relationship with the customer; this statement holds true for crisis as well. Why is maintenance so important? First, it allows building a productive, long-term relationship with the customer, as this is the stage that actually is the lifecycle phase of commercial operation, which, as a rule, is sufficiently long, and typically lasts for several years. It is the stage of maintenance when the customer pays back to the developer for the added product value. Another important aspect is that the maintenance stage makes it possible for the developer to switch to reusable software production. That is, the software developed for a specific customer, if well maintained, after certain improvement may be proposed as a product for the other customers in order to satisfy their business requirements. The better is maintenance quality, the more probable it is that the new product release may satisfy a certain number of customers, and that in the future it may become a COTS product. In this respect, the maintenance stage is critical even for small products, and even in crisis.

The next step of the software product lifecycle is retirement. It usually happens after the software product has served to the customer for a long time and the users are familiar with it—the processes were established at the client's site, and the procedures are thoroughly documented. The users understand the software well; they have a working knowledge to operate it.

However, in certain cases the customers arrive to a conclusion that the product retirement is required. Why does this happen? The answer is that the software, in contrast for, say, material architectural structures, becomes outdated rather fast. The reason is rapidly changing software and hardware environments, so that at some point of time the customers cannot afford the costly maintenance of a legacy software solution. In this case, the customers have to move to a new software product that supports entirely new functionality, the implementation of which is no longer feasible for the currently operating product release. Moreover, if certain functions and data of the legacy software product are still required, it is important to migrate the data to the new application. The retirement process is often non-straightforward and challenging. In terms of human factor, it is often undesirable, especially in case of the large-scale systems with a large number of users and a large size of critical data. Retirement should be based on a well-informed decision, which includes a thorough evaluation. The cost of software replacement includes a number of factors: the cost of technological change, the cost of development and maintenance of the applications based on the new software, training costs, and employee productivity decrease during the transition period.

We have briefly covered all stages of the software development lifecycle. A very important project document in respect to the lifecycle is the project plan. The project plan usually embraces the following phases: requirement analysis and specification, preliminary and detailed design, implementation, testing, integration, acceptance testing, transfer, maintenance, and retirement. The project plan also includes a high-level estimation of such key dimensions of the project as time and budget, typically in terms of project schedule, which contains the main activities and milestones, i.e. the key control points where certain results are achieved. An important concept related to a milestone is a deliverable, i.e. practical outcome produced on reaching each milestone. In addition, the project plan includes a number of lower-level plans, such as a risk management plan, a test plan, an integration plan, and some others.

Depending on the specific model, a software product lifecycle may have certain features. For example, design specification, i.e. an outline of the product components and connections between them, may not be fully detailed, since the degree of details is dependent on the specific model. In some models, there is a lifecycle when the system is operational after one single pass of all the lifecycle stages. In other lifecycle models, the sequential change of phases is performed iteratively, which looks like a cyclic repetition of the stages of lifecycle; each iteration yields to a product with added functionality.

The classical approach to software development is based on the structural analysis and design and is often referred to as structural analysis and development. This approach does not take into account certain aspects, such as dynamics of the product design; this results in the ability to select the type of the programming language for the product only after the product specifications are complete. However, we can overcome this drawback by using a different lifecycle model, such as object-oriented approach. In addition, with structural analysis and development it is often a challenge to implement a large-scale code reuse; however, this is possible with the object-oriented lifecycle model. Since code reuse is among the most important goals of the software lifecycle management, the structural analysis and design approach is not a good recommendation for crisis software development. The problem here is that the deviations in reuse of the product artifacts, which include not only the code but also documentation usually amount to almost 50 % of the project cost. This is a significant source of savings in terms of time and human labor; it is especially valuable in crisis. However, it is a challenge, since such a strategic reuse requires a disciplined development, and the proper use of specific standards. Developers should strive for this, and certain lifecycle models assist them in this respect. One more important feature of the software lifecycle is that the boundaries of the phases of lifecycle can vary and even overlap, which adds agility to crisis management in case of rapid requirement changes. One example of such agility is the object-oriented lifecycle model; however, the payment for such an agility is a high level of discipline in software development.

Let us overview the contribution of the various phases of software products lifecycle in terms of time and cost. The data are based on a number of software projects, which were completed by Hewlett Packard and other large-scale developers [7]. Clearly, maintenance accounts for the lion's share of the projects cost and schedule. However, certain stages, such as coding, taken together with testing and requirement analysis amount to a relatively small share of the product cost. The fact that maintenance amounts to approximately 60 % of the total project cost is especially important for large-scale and mission-critical projects. These projects are usually long-term and include a large number of components to be integrated and maintained as a system, which causes additional risks in a crisis. In addition, the software products developed for maintainability (including portability, expandability and some other quality attributes) tend to be more efficient in crisis than the products, which improved the code by refactoring at the implementation stage. Another interesting finding is that the phases before and after the coding amount to nearly 30 % of the total cost, while the coding itself is only 5 %. Thus, programming itself, no matter how large the project is, is usually inexpensive, and the crisis does not seriously affect it. However, the stages that embrace coding, i.e. design and testing, usually provide a significant improvement in quality. Thus, good design and testing are very important in crisis management since they not only provide better quality but also save time in coding and the later stages.

Most of the serious defects found in software products occur at the early stages of requirement analysis and design specifications. Therefore, these defects are very expensive to fix, because all the later lifecycle activities (including detailed design, coding, testing and integration) have to be repeated, and the new document artifacts to be produced. To detect these defects, a number of methods of analysis exist, including design reviews and formal logic-based verification. We highly recommend to use these methods in crisis conditions. There are also special computer-aided tools to analyze, detect and fix such defects. The cost of detecting and fixing software defects grows exponentially in the lifecycle progress. For example, in case a defect is detected at an early stage of requirement analysis, it is quite cheap to fix. However, if it is detected at one of the later stages, specifically at the stage of maintenance, the fix cost is much higher, because it is now required to change the entire release of the software product including the accompanying documentation.

Each phase of the software lifecycle includes three key components—processes, methods and tools. Under the process, we imply a sequence of the tasks to implement, they are clearly different, i.e. they have a clear entry and exit criteria. Under the method, we imply a relatively formal description of each task in the process. Under the tool, we imply computer-aided software, which supports the software development processes and methods.

In the next chapter, we will discuss in more detail the major types of lifecycle models. These models are: build-and-fix, waterfall, rapid prototyping, incremental, synchronize and stabilize, spiral, and object-oriented.

Build-and-fix model is, in fact, close to trial and error approach; its lifecycle is simplified. Waterfall model is strictly document-driven; it requires single pass of the lifecycle to build a completely operational software. Rapid prototyping is typically combined with other models. Incremental model implies several sequential releases, which add up functionality. Synchronize and stabilize model is aimed at rapid and early testing to maximize return on investment (ROI). Spiral model focuses on risk assessment. Object-oriented model has overlapping phases with intensive parallelism.

What are the common features of these lifecycle models? Typically, they all except the build-and-fix include every of the above-mentioned lifecycle phases or stages. Also, they all usually include a few iterations of these product lifecycle phases, with the exception of the waterfall model. Typically, the lifecycle stages are clearly separated; however, they may go concurrently in the object-oriented model.

Proper application of any model lifecycle requires a high level of the organizational maturity and project team discipline in terms of standards for documentation, coding, computer-aided development with special software tools, and other activities. If the maturity level is insufficient, certain models, such as spiral and object-oriented can degenerate into build-and-fix, that is, the benefits of the model may not work, and project costs may increase dramatically.

There is no "silver bullet" or a universal lifecycle model. Instead, the best lifecycle model is always determined by the scope and scale of the project. Each model has its advantages and disadvantages; we are going to discuss them in more detail in the next chapter.

What does the choice of the software product lifecycle depend on? Primarily, it depends on the nature and size of the product. In this respect, analysis and specification of requirements and constraints for the basic product scope define the selection. The key constraints are the product size, development time and project risks. For instance, spiral model is strongly dependent on risk assessment, so it makes sense to apply it in case risk analysis is required. Lifecycle model choice influences the economics of the project, including return on investment. If it is not critical to apply a full lifecycle model, for example, if incomplete documentation will suffice, it is recommended to save resources for certain steps, such as low-level design. Maintainability is also dependent on the model choice: certain models provide better maintainable products. Lifecycle model choice also determines the progress of development in terms of meeting future customer needs. In addition, the lifecycle model choice determines upgrade path of the product in terms of its evolutionary or revolutionary development. In other words, the lifecycle model determines whether radical changes or constraints for the product architecture are required, or the project is going to evolve gradually. Lifecycle model type determines the speed of defect detection and fixing; for example, the synchronize and stabilize model is aimed at frequent and early testing. Certain models, such as spiral, promote risk management; the others use prototyping for risk mitigation. However, the prototype is usually far from operational product in terms of quality attributes; we are going to discuss the differences between a prototype and a product in the next chapter.

Concerning specific lifecycle models and their features, what is worth mentioning? The build-and-fix model has an incomplete lifecycle, it is suitable for small projects below 1,000 lines of code (or 1 KLOC); it is totally unsuitable for large and complex projects with a high evolution potential. The waterfall model provides intensive feedback at the early stages of the lifecycle, as every stage is completed only after the development of the documents, which allow moving on to the next stage. In the waterfall, it is impossible to start the next stage without having these documents complete, and terminating the previous stage. Rapid prototyping is a dependent model, since it does not yield to an operational product in terms of performance, reliability, security and other quality attributes. Conversely, the incremental model always provides an operational product after each iteration, even though this product is not fully functional. Synchronize and stabilize model aims are early detection of errors and early ROI. Spiral model involves multiple iterations and focuses on risk analysis. Object-oriented model offers iterative design with significant overlapping and high concurrency of the phases.

Let us discuss the advantages and disadvantages of the lifecycle models that we have introduced.

The build-and-fix model is applicable for small products, which do not require complex maintenance; however, it is not suitable for large or medium-scale non-trivial products of over 1 KLOC size.

The waterfall model is document-driven, since the documents mark the completion of each stage, and it provides clear and disciplined product development processes. However, because this model is single-pass only, the resulting product may not meet the requirements of the client.

Rapid prototyping model has a temptation to reuse the code, which did not go through sufficient testing and documenting processes, and which, therefore, must be re-implemented. However, this model assists for requirements analysis and identification of the features, which are most important to the customer.

The incremental model promotes maintainability, because it provides smooth transition from one software product release to another. This model facilitates early return on investment. However, it requires an open architecture that supports an evolutionary improvement of the product; without such an architecture, the incremental lifecycle may degenerate into a build-and-fix.

The synchronize and stabilize model meets the future needs of the customer and provides a high degree of component integration; however, it is rather complex, as it requires intensive testing, specific processes and computer-aided tools.

The spiral model combines a number of features of the above models; however, we recommend to use it for in-house development, as it requires a thorough risk analysis, and it is doubtful that a number of requirements and constraints related to the key risks will be revealed to the outside developers.

The object-oriented model requires development discipline, otherwise it can easily degenerate into the trial and error approach; it features iterative development, intensive interaction and significant overlap between the phases.

The key factors influenced by the choice of lifecycle model are time to market, operational quality and business value of the product, change management and risk management strategy, and customer relations during maintenance. All these factors are mission-critical in crisis.

In case of crisis development, it is also very important to choose adequate tools that assist in software engineering and software product development. These are called computer-aided software engineering (CASE) tools. Software development has a number of aspects. Software in the small can be regarded as the art of programming or development of separate modules, the individual pieces of code. Software in the large can be understood as software engineering; this is a technological approach to software development based on a well-justified choice of the above-mentioned lifecycle models. One more aspect is software as a teamwork, including team development support, which is very important for mission-critical and large-scale software systems that often involve a number of cooperating teams.

The CASE tools help in all three aspects: personal software process improvement, lifecycle optimization, and resource management for team-based development.

CASE tools can be divided into front-end and back-end ones; the former assist in earlier lifecycle stages, while the latter assist in implementation and the post-production. Some CASE tools, such as Rational product line, are implemented as conveyor systems, where each product is responsible for a specific lifecycle stage. Microsoft Visual Studio represents the other approach, which has a common interface for a number of lifecycle operations, such as testing, integration, coding, designing, documenting and so on.

CASE tools provide a distinct advantage for software systems production, specifically on the large and mass scale. However, in practice, they require organizational maturity of the team and working knowledge of the software development standards. In crisis, CASE tools are usually feasible for large-scale projects. For smaller projects, the cost of licenses for the CASE tools and developer training may sometimes be unaffordable. In case of successful application of CASE tools, the team typically achieves a significant productivity growth and a substantial reduction in terms of time and budget.

In crisis, CASE tools usually pay off. However, to estimate their feasibility, we need certain metrics. We also need certain global metrics for the project itself. Which metrics are better? For the high-level project planning, we often use such global parameters as time, budget and functionality. In crisis, however, it also makes sense to do cost-benefit analysis, i.e. to estimate possible benefits for the customer and the developer depending on their investments. For certain lifecycle stages, such as testing and maintenance, we can use specific metrics. Every lifecycle stage has its own metrics. The testing stage, for instance, can use such metrics as complexity of a separate module, number of lines (typically measured in KLOC), number of different operators in a module, relative number of errors detected for each KLOC and so on. It is required to analyze the total number of failures and identify their lifecycle phases in order to manage detection and removal of the defects injected prior to testing. Naturally, not all of the design errors can be detected in the implementation or integration phase, i.e. prior to the transfer to the customer. Thus, metrics should be used together with defect reports, which include defect status. Furthermore, crisis imposes certain constraints on defect source identification and decision-making depending on severity and persistency; therewith, a number of metrics from the previous lifecycle stages are applicable. The decision is usually human factor-dependent: for example, the project manager decides at which point of time it is feasible to stop the testing and switch to product transfer. In crisis, metrics are still affordable; however, usually simple metrics, such as KLOC, are efficient enough and pay off.

5   Conclusion

We overviewed the concept of crisis and its application to software development and software product lifecycle.

Managing software development lifecycle is a challenge in case of large-scale and mission-critical applications, especially in crisis. To solve the challenge, a uniform methodology for managing the software development lifecycle is required, which includes models, methods and supporting CASE level tools.

We discussed the general lifecycle pattern and its stages, such as requirement analysis and specification, design, implementation, integration, maintenance and retirement. We said that the cost of defect detection and fixing increased expo- nentially as we moved from the earlier stages to the later ones, so error detection should happen as early as possible. There are special techniques for error detection in every lifecycle stage; these include the processes, methods and tools.

Concerning the lifecycle models, the model selection determines the key parameters of the project. Selection of the lifecycle model affects a number of critical parameters of the software development project, and it often determines its overall success. The most essential of these parameters are the architecture of the project, its budget and timeframe. Model selection also determines a number of required and optional project artifacts and certain quality attributes of a software product, so that it is possible to decrease the product time to market. The above-mentioned parameters are based on product quality metrics, which make crisis management more accurate and predictable.

The lifecycle model selection should be adequate to the experience of the project team in terms of problem domain expertise and operational knowledge of specific technologies, CASE tools and documenting standards. We briefly discussed a number of lifecycle models, such as build-and-fix, waterfall, incremental, object-oriented, spiral, and a few others. Some of the models considered require iterative development; others are more straightforward and document-driven. Certain models, such as spiral or object-oriented, require a high level of discipline and organizational maturity; otherwise the lifecycle can easily degenerate into trial and error approach.

There is no "silver bullet", i.e. no universal model, which suits any software product equally well. The scope and size of the project are the determinants of lifecycle model selection, and we can customize the models in order to adjust for the specific features of the project. For some models, such as rapid prototyping, we recommend a combination with the others. We identified the key advantages and disadvantages of each model discussed; the next chapter will present a more detailed discussion.

References

1. Software Engineering Institute: Retrieved November 25, 2015 from www.sei.cmu.edu/
2. Sommerville, I.: Software Engineering, 864 pp. 8th edn. Addison Wesley (2006)
3. Systems and Software Quality Requirements and Evaluation: Retrieved November 25, 2015 from http://www.iso.org/iso/home/store/catalogue_ics/catalogue_detail_ics.htm?csnumber=35733
4. Barendregt, H.P.: The lambda calculus (rev. ed.), Studies in Logic, vol. 103. North Holland, Amsterdam (1984)
5. Curry, H.B., Feys, R.: Combinatory Logic, vol. 1. North Holland, Amsterdam (1958)
6. Roussopulos, N.D.: A Semantic Network Model of Databases. Toronto University (1976)
7. Schach, S.R.: Object-Oriented and Classical Software, 688 pp., 8th edn. McGraw-Hill (2011)
8. Scott, D.S.: Lectures on a mathematical theory of computations, 148 pp. Oxford University Computing Laboratory Technical Monograph. PRG-19 (1981)
9. Wolfengagen, V.E.: Event driven objects. In: Proceedings of CSIT'99, Moscow, Russia, pp. 88–96 (1999)
10. Wolfengagen, V.E.: Applicative Computing. Its quarks, atoms and molecules, 62 pp. JurInfoR, Moscow (2010)
11. Fowler, M.: Analysis Patterns: Reusable Object Models, 223 pp. Addison Wesley (1997)
12. Kalinichenko, L., Stupnikov, S.: Heterogeneous information model unification as a pre-requisite to resource schema mapping. In: ITAIS 2009, pp. 373–380. Springer (2009)
13. Zykov, S.V.: Enterprise content management: bridging the academia and industry gap. In: Proceedings of i-Society 2007, Merrillville, Indiana, USA, vol. I, pp. 145–152, 7–11 Oct 2007
14. Zykov, S.V.: Integrated methodology for internet-based enterprise software systems development. In: Proceedings of WEBIST2005, Miami, FL, USA, pp. 168–175, May 2005
15. Zykov, S.V.: An integral approach to enterprise content management. In: Callaos, N., Lesso, W., Zinn, C.D., Zmazek, B. (eds.) Proceedings of the 11th International World Multi-Conference on Systemics, Cybernetics and Informatics (WMSCI 2007), Orlando, FL, USA, vol. I, pp. 212–216, 8–11 July 2007
16. Zykov, S.V.: The integrated methodology for enterprise content management. In: Proceedings of the 13th International World Multi-Conference on Systemics, Cybernetics and Informatics (WMSCI 2009), Orlando, FL, USA, pp. 259–264, 10–13 July 2009
17. Guha, R., Lenat, D.: Building Large Knowledge-Based Systems: Representation and Inference in the Cyc Project. Addison-Wesley (1990)
18. Evans, E.: Domain-Driven Design: Tackling Complexity in the Heart of Software, 560 pp. Addison Wesley (2003)
19. Zykov, S.V.: ConceptModeller: a frame-based toolkit for modeling complex software applications. In: Baralt, J., Callaos, N., Chu, H.-W., Savoie, M.J., Zinn, C.D. (eds.) Proceedings of the International Multi-Conferences on Complexity, Informatics and Cybernetics (IMCIC 2010), Orlando, FL, USA, vol. I, pp. 468–473, 6–9 April 2010
20. Zykov, S.: Pattern development technology for heterogeneous enterprise software systems. J. Commun. Comput. 7(4), 56–61 (2010)
21. Naur, P., Randell, B. (ed.): Software Engineering: Report on a Conference sponsored by the NATO Science Committee, Garmisch, Germany, 7th to 11th October 1968, Brussels, Scientific Affairs Division, NATO, 231 pp., January 1969
22. Randell, B.: The 1968/69 NATO Software Engineering Reports. Dagstuhl-Seminar 9635: "History of Software Engineering". Schloss Dagstuhl, 26–30 Aug 1996. Retrieved November 25, 2015 from http://homepages.cs.ncl.ac.uk/brian.randell/NATO/NATOReports/index.html
23. Naur, P., Randell, B. (eds.): Software Engineering: Report on a Conference sponsored by the NATO Science Committee, Garmisch, Germany, 7th to 11th October 1968, Brussels, Scientific Affairs Division, NATO, 231 pp, January 1969
24. Randell, B.: Software engineering in 1968. In: Proceedings of the 4th International Conference on Software Engineering, pp. 1–10, Munich (1979)

25. Dijkstra, E.: The humble programmer. ACM Turing Lecture, Comm. ACM **15**(10), 859–866 (1972)
26. Gray, G.T., Smith, R.Q.: After the B5000: Burroughs third-generation computers 1964. IEEE Ann. Hist. Comput. **31**(2), 44–55 (1980)
27. MacKenzie, D.: Mechanizing Proof: Computing, Risk, and Trust, 440 pp. MIT Press (2004)
28. Bourque, P., Fairley, R.E. (eds.): Guide to the Software Engineering Body of Knowledge, Version 3.0. IEEE Computer Society (2014). Retrieved November 25, 2015 from http://www.computer.org/web/swebok/v3
29. The Carnegie Mellon University History: Retrieved November 25, 2015 from http://www.cmu.edu/about/history
30. Ghezzi, C., Jazayeri, M., Mandrioli, D.: Fundamentals of Software Engineering, 624 pp, 2nd edn. Pearson (2003)
31. Lipaev, V.V.: Software Engineering. Methodological Foundations, 680 pp. TEIS, Moscow (2006) (in Russian)
32. Marx, K.: Capital: a critique of political economy. Volume II. In: Engels, F. (ed.) Book One: The Process of Circulation of Capital. Meissner, Hamburg (1885) (In German)
33. Hardy, E.: The economic crisis—the Marxian explanation. World Socialist No.1, pp. 20–26, April 1984
34. Bautsch, M.: Cycles of software crises. In: ENISA Quarterly on Secure Software, vol. 3, no. 4, pp. 3–5, Dec 2007. Retrieved November 25, 2015 from https://www.enisa.europa.eu/publications/eqr-archive/issues/eqr-q4-2007-vol.-3-no.-4/at_download/issue
35. Deming, W.E.: Out of the Crisis. MIT Center for Advanced Engineering Study, Cambridge, MA (1986)
36. Lenat, D., Reed, S.: Mapping Ontologies into Cyc, AAAI 2002 Conference Workshop on Ontologies for the Semantic Web. Edmonton, Canada(2002)
37. Birnbaum, L., Forbus, K. et al.: Combining analogy, intelligent information retrieval, and knowledge integration for analysis: a preliminary report.In: ICIA 2005, McLean, USA (2005)

Chapter 2
Software Product Lifecycles: What Can Be Optimized and How?

Abstract The chapter discusses lifecycle models for software development in more detail. These include build-and-fix, waterfall, incremental, object-oriented and spiral. We present a more detailed description of the lifecycle models application for software development. We compare benefits and shortcomings of the models discussed. We confirm that there is no "silver bullet", i.e. a universal lifecycle model equally applicable to any software product. Consequently, lifecycle model choice is dependent upon product size and scope; each project requires a unique combination of features. In crisis, we recommend to combine prototyping with the other models that we discussed in order to achieve a common understanding of the key product features and to reduce project risks. The lifecycle model choice determines project economics, time to market, product quality and overall project success. However, the product success essentially depends on human factors, which include common vision of the critical product functions, transparent communication and feedback. We analyze applicability of the lifecycle models to large-scale, mission-critical software systems, which is essential in crisis. Finally, we introduce a methodology, which includes a spiral-like lifecycle and a set of formal models and visual tools for software product development. The methodology helps to optimize the software product lifecycle, which is mission-critical in crisis. The methodology is applicable to large-scale, complex software products for heterogeneous environments.

Keywords Software lifecycle · Lifecycle model · Software development methodology

1 Introduction

The previous chapter gave a brief review of a number of lifecycle models used in software development, such as build-and-fix, waterfall, incremental, object-oriented, spiral, and a few others.

This chapter presents a more detailed description of the lifecycle models application to software development. It includes discussion of their benefits and

© Springer International Publishing Switzerland 2016
S.V. Zykov, *Crisis Management for Software Development
and Knowledge Transfer*, Smart Innovation, Systems and Technologies 61,
DOI 10.1007/978-3-319-42966-3_2

shortcomings. It analyses applicability of the lifecycle models to large-scale, mission-critical software systems, especially in a crisis.

Some of the models are more straightforward, others require a number of iterations. Our deeper investigation of the models will still conclude that there is no crisis-proof "silver bullet" for lifecycle models. However, we will arrive to certain recommendations of combining and adjusting the models in order to succeed in crisis software development.

Project success is usually determined not only by the lifecycle model, or by a combination of models, but also by a number of human factors, which may help or hinder a common understanding of the key product features by the client and the developer. We will cover these human-related factors in more detail in Chap. 5.

In order to optimize software product lifecycle, which is mission-critical in crisis, we will introduce a methodology that includes a spiral-like lifecycle and a set of formal models and visual computer-aided tools for software product development.

This chapter is organized as follows. Section 1 discusses the abbreviated and straightforward lifecycle models, such as build-and-fix and waterfall. Section 2 presents an overview of simple iterative models, such as incremental and prototyping. Section 3 describes more complex software lifecycle models; these are spiral, synchronize and stabilize, and object-oriented. Sections 4 and 5 contain an overview of an enhanced software development methodology, which provides lifecycle optimization and sequentially elaborates the deliverables for mission-critical software products in crisis. The conclusion summarizes the results of the chapter.

Let us have a look at the software development lifecycles in more detail.

2 Simple Lifecycles: Brief and Straightforward

One of the models of software development lifecycle discussed previously is the build-and-fix (see Fig. 1). This is a model of incomplete lifecycle. Because of its simplicity, the build-and-fix is not suitable for large and complex projects, which have a size of over 1 KLOC. So, the build-and-fix model may be a possible option for a software solution, which is downsized by crisis. However, it is only applicable in case of a trivial product with clear requirements.

The other model discussed previously is rapid prototyping (see Figs. 4 and 5). It is also somewhat limited, despite the fact that it includes all the necessary stages of the lifecycle. These are analysis and specification of requirements, preliminary and detailed design, implementation, unit testing, integration, product testing, maintenance, and retirement. The limit of the rapid prototyping is lack of self-consistency. Actually, its testing phase, both for the individual modules and the prototype as a whole, yields to a low quality code. The prototype documentation is usually insufficient and incomplete, and the resulting code is not a software product, since it only simulates the key functionality and certain aspects of the future software system of operational quality.

Fig. 1 Build-and-fix model

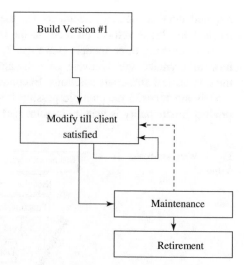

Fig. 2 Waterfall model

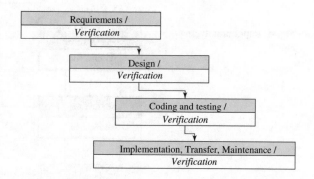

The waterfall model presented in Fig. 2 is fully applicable to large-scale and mission-critical software systems; however, it has some limitations, since it has a limited agility to meet the crisis conditions. In particular, the waterfall model requires discipline and organization, operational knowledge of CASE tools as the project team needs to produce a large number of documents and to communicate intensively. The documentation should meet the standards, which follow the agreement with the customer. Any product document, such as operations manual, should follow specific templates, since documentation is a critically important part of any waterfall-based software product. Let us recall that the product is not only the code but also a large amount of documentation required for competent and stable maintenance. The product documentation has a special value for the maintenance personnel, as they usually read the code produced by other developers, and their task is to detect and to fix the remaining defects. The documentation is mission-critical for the waterfall model, because developers follow the lifecycle based on document-driven milestones conditions. For example, as soon as the required detailed design documents for the product are ready, the milestone is reached and the developer proceeds to the implementation phase of the lifecycle.

The following three models (see Figs. 3, 4 and 5) are important to understand how to organize the lifecycle of software systems, including large-scale and mission-critical anti-crisis solutions. In contrast to the waterfall pattern, these three models are focused on multiple passing through the stages of the lifecycle. The product functionality is usually incremented after each pass.

Fig. 3 Waterfall-based V-model

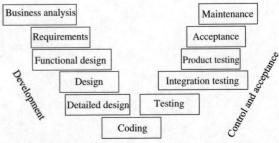

Fig. 4 Rapid prototyping model

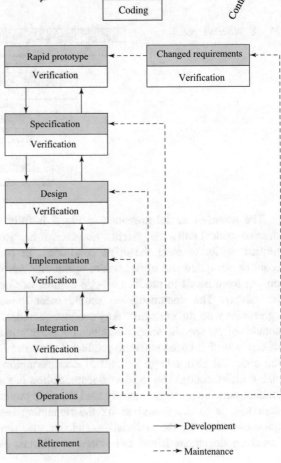

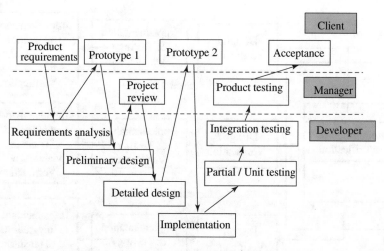

Fig. 5   Rapid prototyping-based "shark teeth" model

Naturally, it is challenging to implement certain kinds of the software systems in a single pass. However, the waterfall is adequate for a problem domain with clear and stable requirements, which is straightforward to document and design. The waterfall approach is applicable mostly for government agencies and military software applications.

3   Simple Iterative Lifecycles: Incremental and Prototyping

The iterative lifecycle models provide loop-based elaboration of the software product. They assume that several loops are required in order to build a product release. Each of these loops usually includes all stages of the software product lifecycle. The iterative models are often easier to follow in terms of discipline, as they do not usually require developing a full product functionality or a complete product documentation after each stage.

One of these iterative models is the incremental model (see Fig. 6). What are its key features? According to the model, while building a project plan, the product is divided into a sequence of releases. The lifecycle stages, which precede product transfer to client, involve a number of releases. These are iterations of the development cycle, each of which provides an operational product. Therewith, in case the product does not require a revolutionary transformation of the previous releases (i.e. functionality builds up smoothly), each release is transferred to the customer as an operational product, though it has a limited functionality. In addition, every lifecycle stage delivers the required product documentation, so that each product release is operational and utilizable.

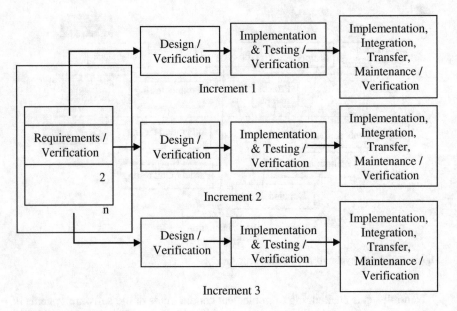

Fig. 6 Incremental model

For instance, in case of an online store, we can initially simplify the interface associated with the purchase of products. The first release may not have a choice of delivery options (e.g. by sea and by air), and it may have a single kind of delivery with a fixed rate. Later releases can also include more details to support credit card payments, such as a dedicated server for client authentication and transaction processing. However, even the first release yields to a fully operational product, though it is rather simple in terms of functions available.

Thus, the idea of the incremental lifecycle is to supply an operational product to customer as soon as possible. In some cases, this can be a suitable solution in terms of crisis management. For the incremental model, the project plan typically specifies the sequence and schedule of functionality transfer to the customer; it may also include a maintenance plan, which specifies technological and functional constraints for each release.

Another feature of the incremental model is a relatively smooth transfer of new functionality. Each release is a clearly separated functional block, and it contains a number of modules. Naturally, these modules do not exist by themselves and are related to some other modules. They can inherit certain properties of these other modules; they also can interact with semantically related modules through the interfaces provided. Thus, it is desirable that within every release, each interacting module is relatively small and self-consistent, i.e. it has a relatively small amount of interaction points with the other modules. By keeping the modules and the releases relatively small and self-consistent, the incremental lifecycle provides a smooth transfer of the new functionality to the customer.

Modular software design ensures minimum connectivity between the modules, so that each relatively small and functionally separate task is located in a separate software module. The same modularity principle usually holds true for each of the incremental releases. However, since the functionality is implemented and introduced gradually, the new modules and releases will interact with the existing ones, so we need to test their interfaces. Therefore, if the product requires a revolutionary functional change, which significantly influences its previous releases, it usually causes a number of problems. Thus, we can get a local crisis in development instead of a stable incremental release plan.

The higher-level source of such a local crisis can be poor design and inadequate planning. As for the lower-level sources, we can identify at least two of them at this point. First, there is a significant problem associated with inheritance. It may happen that a number of modules in the operating release is to change in a significant way. Therewith, we have to redesign and rebuild a significant percentage of the previous release structure, and to rewrite all the documentation required. Of course, any new release brings certain changes to the previously built modular structure. However, with one revolutionary functional update, we have to make such a large number of changes in design and implementation, that it nearly nullifies all the efforts to produce the previous releases. In fact, this local development crisis is comparable to complete redevelopment of the product from scratch in a build-and-fix manner. In this case, the functionality developed for the previous releases would be largely rebuilt, and the time and labor to create this functionality would be lost. In this respect, revolutionary development typically results in a local crisis of an incremental lifecycle.

An incrementally developed product should have a scalable architecture in terms of release updates. For example, a web service-based component architecture usually scales up well in terms of adding new modules or expanding existing ones. However, there are software architectures, such as a file server, that do not support similar scalability equally well. Therefore, we should consider the features of a particular model at the early stages of project planning and high-level architectural design in order to adapt to the technical constraints and to avoid a local crisis in product development.

For the developers, the incremental model provides evolutionary interaction with the customer and greatly simplifies their relations, because the core modules that implement the business logic of the application often vary slightly. The new releases only add functionality. Therewith, maintenance of an incremental product is usually sufficiently smooth and relatively inexpensive.

However, a possible disadvantage of the incremental model is that it is not suitable for quite a number of software products, which initially require a full-featured implementation. Let us assume that there is a number of customers, who need a full-featured online store, which includes a 3D catalog, credit card payment support, and a variety of electronic payment gateways. Certain clients would also ask to monitor delivery, as it is implemented at their competitor portals, and it is convenient and useful. If the product initially requires full functionality, we should probably consider some other model, such as waterfall, which allows a single pass implementation. Of course, in case of waterfall certain project risks are higher; however, they will be discussed further in relation with the spiral model, which is designed to deal with them. Therefore, there is a number of constraints for the incremental model, and it is clear that this model it is not suitable for every product.

Another drawback of the incremental model is that the resulting software product should provide a stable upgrade path for its development. That is, the efforts spent for functional updates with each product release should clearly exceed the redundant efforts for the high-level release reconfiguration. Such reconfiguration efforts should not have a significant negative impact on the performance of the next product release. The incremental model does not support a revolutionary, unstable path of the software upgrade; it also has no mechanisms for risk assessment.

Depending on customer or market constraints, a number of software products requires revolutionary changes in the product concept itself, such as fundamental principles that underlay the functional requirements, project plan and product release policy. If the customer requirement changes are frequent, spontaneous and dramatic, and there is no way to adjust these requirements so that they become evolutionary, it may turn out that every other release the developer has to create a new product almost from scratch rather than to reuse a significant portion of the previous one. Thus, the incremental approach degrades to build-and-fix. Moreover, in contrast to build-and-fix, which is an incomplete lifecycle model, the developer has to re-implement the entire lifecycle for each release. For each release, the developer has to specify complete requirements, to do the software design, i.e. to produce a large number of diagrams, including data flow diagrams, use cases, class diagrams etc. The developer also has to develop a new test plan, including product testing scenarios and their sequence, acceptance test cases and a number of other artifacts. Moreover, the end user and administrator documentation requires significant changes. The new product probably has a different setup procedure, user interfaces, usage scenarios, error codes and so on. The developer has either to rebuild all these documentation artifacts or to create new ones. Thus, the developer has to rework the product using a more bulky and complex approach than a trivial build-and-fix, which includes a full-scale documentation and artifact reviews for each software lifecycle phase. Therefore, the software production is likely to result in a local crisis, since it involves a huge amount of bulky overheads. Thus, the incremental model is unacceptable for a product that quickly goes beyond the original concept, no matter how large the product is.

In case of a predictable upgrade path of the product, the previous release is naturally included into the next one. At the same time, a special document, release notes, is issued, which includes a list of additions to the previous release. Release notes document also contains important information about the new release of the software product. It addresses customer's maintenance service, who detect, localize and fix errors; it also guides customer's end users on their moving from the previous release to the next one.

Figure 6 shows a view of the incremental lifecycle model. It is clear that each subsequent release includes the functionality of all the previous ones. Thus,

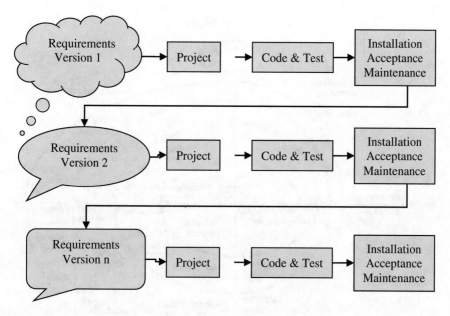

Fig. 7 Evolutionary model

functionality increases smoothly, and so that each of the following releases absorbs previous ones and adds certain new features. For incremental product development, production of the new release includes verification of all the lifecycle stages, such as requirements analysis, requirements specification, design etc. Thus, the main stages of the software lifecycle are the same for each release. The incremental model fits evolutionary introduction of product functionality.

Figure 7 shows a specific form of incremental development model, which is called evolutionary. It provides a gradual transition from the previous release to the next one; each release elaborates functionality rather than builds it up. The rest lifecycle processes are similar to the incremental model.

4   Complex Iterative Lifecycles: Spiral, Synch-and-Stabilize and Object-Oriented

Another iterative approach to software systems lifecycle is the so-called spiral model introduced by Boehm [1]. According to the approach, each iteration consists of four phases (Fig. 8):

(1) determine the goals for product and business objectives, understand the constraints, suggest possible alternatives;
(2) evaluate the alternatives by risk analysis and prototyping;

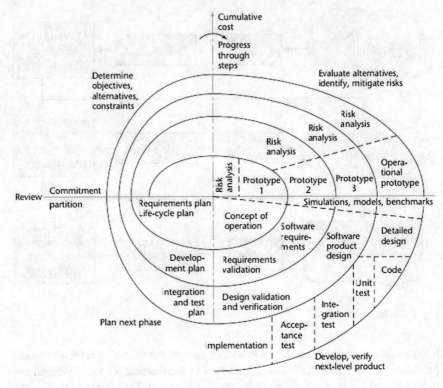

Fig. 8 Spiral model (© 1988 IEEE)

(3) develop a product by detailed design, coding, unit testing and integration;
(4) plan for the next iteration, including product development, implementation and
     delivery to customer.

The above four stages: determine—evaluate—develop—plan, are often represented graphically as a spiral.

This model is suitable for projects with significant risks. Some other lifecycle
models also address risk assessment. In crisis, the projects tend to become more
risky. Additional crisis-related risks may include delays of funding, communication
challenges in the project team, especially in case it is distributed. In fact, in the
spiral model, risk analysis happens each iteration.

Each phase of the spiral model usually repeats; there is often three to four
iterations. However, the exact number strongly depends on the "convergence" of the
project. In certain cases, the number of iterations is difficult to predict; it may also
happen that after the risk assessment it is not feasible to continue the project. This
may result in additional expenses; however, in crisis conditions it is required to
recognize that the project team is not able to ship the fully functional product of the

required quality level within the deadlines. A possible solution of this problem is the contradiction management matrix-based approach; we discuss it in the next chapter.

Each cycle includes four basic phases: determine, evaluate, develop and plan.

The first phase includes an outline of the objectives for the current iteration, possible alternatives to achieve these objectives, and the constraints for each alternative. Further, evaluation of the alternatives follows; risk assessment is among the key activities. Risk assessment is a complex process; it requires specific knowledge. In crisis, risk experts often have to make decisions in case of uncertainty, insufficient resources and incomplete information. Prototyping helps to reduce some of the risks and to understand the others better; this chapter discusses prototyping in more detail below. After risks identification, risk mitigation plan follows, which specifies the ways to reduce risk consequences or to continue the project with the risks that exist. Then, the implementation phase begins, which starts from coding and testing of the functions required in the iteration, and which ends with the integration and testing of the partial product developed for the current loop of the spiral. Afterwards, based on development postmortem and the existing resources, the next loop of the spiral is planned.

Risk analysis often includes a number of significant uncertainties, which are unlikely disclosed by the customer. Risk analysis usually requires a large amount of expensive labor, so spiral model is generally feasible for large-scale projects. The spiral model is suitable for the so-called in-house projects, where the developer and the customer collaborate within same enterprise. Typically, in large corporations, there is a dedicated IT company, such as Gazprom Inform as a part of Russian Gazprom group of companies. The spiral model is a suitable solution for such enterprises, since the developer and the customer belong to the same large-scale corporation. In case of in-house development, there is usually an adequate transfer of the sensitive information required to assess the risks, and risk assessment results are reliable. Moreover, the in-house development with spiral model is cost-effective, and so this is a recommended option for crisis software development.

The spiral model is somewhat similar to iterative models, such as incremental and evolutionary. However, spiral model is fundamentally different from a number of other models because of explicit risk assessment. Certain lifecycle models can be combined with the others, especially with the rapid prototyping. Spiral model also includes prototyping, which usually assists in risk analysis and evaluation. Rapid prototyping helps a developer to discuss possible product alternatives with the customer. As compared to a full-scale software product, a prototype is relatively cheap and easy to produce. A prototype behavior is usually functionally similar to the product; however, it is limited in terms of quality attributes, such as performance, reliability, security and so on, and in terms of documentation. In crisis, we recommend to combine every lifecycle stage with rapid prototyping, including the early stages, such as analysis and design. Prototyping is a quick and a low-cost way to mitigate a number of project risks. Rapid prototyping simplifies decision-making prior to the release production, so the product transfer occurs timely even in crisis conditions, though the functionality maybe somewhat limited.

The spiral model requires risk analysis; it identifies and classifies the project risks. Developers need to mitigate the most serious project risks, i.e. they have to find a way to reduce their impact on project schedule, budget and product functionality. In case it is impossible to mitigate critical risks, the project manager may decide to terminate the project.

What are the advantages of the spiral model? First, it ensures a smooth transition of the product to the customer. Therewith, it is possible to reuse the product, even under initially high project risks, or in a crisis.

Based on risk analysis, quality assurance metrics are set. Release-based product transfer and risk assessment assist for maintainability. Despite the high costs of risk assessment, the spiral model provides a relatively cost-effective maintenance, which is the most expensive part of the lifecycle. Thus, in terms of full lifecycle the spiral model is often affordable.

The drawbacks of the spiral model originate from high costs of risk assessment. It is applicable for in-house projects.

The spiral model is theoretically applicable to relatively small projects; however, given the substantial costs of the risk assessment, it is more suitable for large-scale ones.

The spiral model requires high level of expertise in risk assessment. In case the development team has no internal risk experts, they have to hire third-party professionals.

The next model we are going to discuss is the synchronize and stabilize model, which is somewhat similar to the Microsoft Solution Framework (MSF) methodology. The next chapter gives a more detailed description of MSF. Due to significant complexity and specific knowledge, skills and CASE tools mastery required, the model is not widespread outside of Microsoft.

This is an iterative model, and the functionality is usually delivered in releases, from essentials to desired requirements, which is similar to incremental model. Each iteration includes planning, design, development, synchronization, integration and stabilization.

According to the model name, the key processes in this software lifecycle are synchronization and stabilization. The synchronization process, however, refers not only to integration of the deliverables produced by the project team, but also to product conformance checking against the requirements specification. The purpose of this phase is to detect and to record as many defects as possible, and to do this as early as possible. However, the model does not suggest immediate correction of the defects recorded in the synchronization phase. Instead, the defects recorded are prioritized by severity and fixing cost, and the list of the defects to fix is produced.

Later on, in the stabilization phase, all the defects detected previously and included into the list are fixed, and the product release for current iteration is produced. The main objective of the stabilization phase is to produce a release with a stable behavior. Therefore, each release is intensively tested in order to meet the threshold values for key quality attributes, such as performance, availability, security and so on. The model uses the idea of sequential functionality build-up;

each release results in an operational software product. The final product usually requires three to four incremental software releases.

Synchronize and stabilize processes are a part of each release. Synchronization process is followed by integration: the individual product modules developed by programmers are assembled in order to make a product release. The integration process is accompanied by frequent and extensive testing, which potentially results in a fast delivery of the product release. The stabilization process ends when all critical errors found in test are fixed.

Thus, synchronize and stabilize are the two interrelated processes, which result in software product release if performed consistently. The final step before release transfer is its "freezing", i.e. saving its configuration.

Let us consider the advantages of the synchronize and stabilize model. First, they come from early and frequent testing. Why is this useful? We mentioned earlier that the defects in the product must be detected as early as possible. The later a defect is detected, the more effort is required to fix it. It may also happen that a defect found in one of the modules affects the operation of the adjacent modules, larger product components, or even the entire product. Furthermore, the defect fixes often crosscut through a number of product artifacts, since they affect not only the code but also the documentation. The documentation is often a crosscutting concern, since it usually influences not only the defective module but also its interaction with the other modules. Of course, it is possible to localize and fix even a logical defect of a top-level module, which is responsible for the overall business logic of the software product. However, such a fix will often influence a large number of the dependent modules and the related documentation. Thus, frequent and early testing is a positive solution and a potential advantage of the synchronize and stabilize model.

However, this advantage often has a side effect: intensive testing may lead to quite a large labor overhead, since it requires specific software, methods and skills. In this case, much time is wasted for synchronize and stabilize processes, which, in fact, do not add any new functionality. Although, in case of proper use, frequent and early testing leads to an exponential increase in product quality, since the number of errors found in testing decreases exponentially; it also provides a better maintainability and customer satisfaction.

Another advantage of the synchronize and stabilize is continuous product interoperability. This is usually guaranteed by partial testing of the product modules at their early development stage. Even before the first stabilization round is over, there exists an operational version of a partial product, which has been thoroughly tested. That is, before the release integration, each combination of potentially interactive modules has already been tested. Continuous product interoperability is vital in case of mission-critical and large-scale systems, which usually combine a huge number of modules that interact in a complex way. For example, the Oracle e-Business Suite, which is an enterprise resource planning system, contains about two dozens of subsystems for planning and management of different kinds of resources: HR, financials, documents and so on. Thus, it is quite challenging to ensure quality and efficiency of such a system without continuous product interoperability.

One more important advantage of this model is that an operational product exists immediately after the initial release. Due to frequent and early testing and continuous interoperability, the initial release is not merely a prototype, but rather an operational quality product with all the documentation required. This results in faster ROI, smoother transfer and better maintainability, which are mission-critical in crisis.

The other possible advantage of the model is that the product becomes potentially better adjustable to the crisis requirement changes. For example, we can adjust the less critical functions and even certain aspects of the low-level architecture by adapting the structure and functions of the modules for the future releases. This may help in future release adjustment, because we can adapt the later releases considering the product shortcomings in terms of architectural design and functionality of the earlier releases.

Additionally, the developer can identify requirement inconsistencies and try to resolve them with the customer in progress of early releases, long before the final release is ready. This approach can significantly reduce the redesign costs for the later releases and can be a positive solution for crisis. Customer's engagement into pre-release testing phase may become an additional source of crisis agility.

The synchronize and stabilize model is flexible and therefore potentially prospective. However, it has a number of complex processes, with the key indicators somewhat difficult to measure and control. Its major drawback is a hardly predictable amount of time for the synchronization and stabilization processes. These processes are designed to add product quality; however, they do not add any new functionality, and in case of immature team, this may result in critical overall performance dropdown.

Under the synchronize and stabilize model, the cycles of integration and testing must take place frequently; in some cases, they occur on a weekly basis. This suggests that every iteration should not only add new functionality but also synchronize and stabilize the intermediate releases, also known as builds. So, frequent build production requires not only new functionality development to match the product specifications but also comprehensive testing of the documentation and code changes with specific methods and CASE tools, in order to detect and fix defects. Therefore, the short intervals between the builds require extremely high productivity and operational knowledge of the methodology in order to be able to add new functions and to test the quality. Otherwise, the developers spend too much time in test, and they have not enough time to add the required functionality to the build. The benefits of this model are often hard to implement, especially in crisis, as they require specialized training and costly staff.

One more lifecycle model to consider, the object-oriented model, is the most dynamic and concurrent. Figure 9 shows the fountain model, a subtype of object-oriented model, which we are going to discuss further.

What are the features of the object-oriented model? The above-mentioned models contain isolated, clearly separated lifecycle stages. These are: requirements analysis, requirements specification, preliminary and detailed design, implementation and unit testing, integration, acceptance testing and maintenance, and

Fig. 9 Fountain
(object-oriented) model

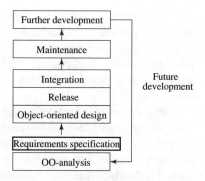

retirement. Waterfall model gives most clear example of this lifecycle stages separation: every next lifecycle stage may start only after the document has been signed, which certifies that the previous lifecycle stage is complete. Conversely, the object-oriented model features intensive interaction between the lifecycle phases. Moreover, there is a significant phase overlap between requirements analysis and requirements specification, and sometimes also between analysis and design, which generally refer to separate phases of the other lifecycle models.

Another important feature of the object-oriented model is its iterative nature. Software product is produced in loops, which often allow returns to the previous lifecycle phases. For example, the phase of object-oriented design often includes a backtrack to the phase of object-oriented analysis. More specifically, analysis of scenarios of product behavior is based on use case diagrams, which are deliverables for product design stage.

Figure 9 shows that the design, analysis and specification phases, as well as design and implementation, are closely related; moreover, returns to the previous phases are possible.

What are the benefits of the object-oriented model? It fits well into the state-of-the-art object-oriented approach to software development, which has been adopted by a large number of industrial programing languages, such as C++, Java and C#. Therefore, object-oriented model is widely used in the production of mission-critical and large-scale systems. This is so because the object-oriented approach allows scalable design of software products due to inheritance and abstraction principles. Based on primitive classes, a small size product can scale up to a large and a complex one.

However, there is a number of features of the object-oriented approach, resulting from principles of inheritance and polymorphism, which may, in case of undisciplined development, lead to local crises in the design and implementation, specifically for large-scale and mission-critical software systems. In particular, concerning the use of inheritance, a bulky and complex class hierarchy may lead to such a situation that, for instance, due to an inaccurate initial problem statement, the system redesign will dramatically modify the entire class hierarchy. This is known as the "fragile" base class problem; it requires complete hierarchy redesign, including code updates for the topmost hierarchy classes that contain the high-level logic and the problem domain-specific features. For complex problem domains, it may occur that the initial design does not scale up, and that a serious redesign of the entire "fragile" hierarchy is required. This redesign usually results in significant labor costs and product delivery delays. In this sense, such a potential benefit of the object-oriented model as inheritance may result is a local software development crisis.

Another potential source of a local crisis is the dynamic method call based on the fundamental object-oriented concept of polymorphism. The object-oriented poly-morphic functions are potentially powerful and resource efficient, as they can uniformly handle heterogeneous parameters. However, these parameters are instantiated only at runtime, which means that it is impossible to test the product for all possible scenarios of the polymorphic function calls. This may result in unpredictable and severe faults that usually lead to critical product malfunctions, such as system crash, data loss, unexpected behavior with system hanging or freezing, and so on.

Thus, the object-oriented model, based on a number of promising concepts, such as inheritance and polymorphism, can degenerate into build-and-fix in large-scale and mission-critical projects, especially under lack of development discipline and organizational maturity. Conversely, well-disciplined and mature development and persistent implementation of standards for coding, testing and documenting usually help to avoid the local crises of the object-oriented model. Therewith, it becomes clear that the root cause of the crisis in software product development is largely dependent upon human-related factors.

5   Managing Lifecycles: Flexible Methodologies

In addition to lifecycle models, there is also a set of lifecycle approaches based on the flexible methodologies such as Agile (Fig. 10), Scrum (Fig. 11), and eXtreme Programming or XP (Fig. 12). Chapter 3 discusses these in more detail. Note that the processes of the software development methodologies are parallel to the phases of the lifecycle models. The methodologies, unlike the lifecycle models, are usually applicable to the projects, which feature greater uncertainty, more risk, i.e. to crisis conditions of software product development. The other aspect of the methodologies is managerial; in addition to practices of software product development they also include a number of project management techniques.

We have discussed a number of lifecycle models—build-and-fix, waterfall, spiral, rapid prototyping, incremental, synchronize and stabilize, and object-oriented—in terms of their applicability for crisis software development. The build-and-fix model is usually suitable for crisis in case of product downsizing, as it works well for small projects with a predictable development lifecycle. The waterfall model is better applicable to large-scale and mission-critical systems. However, waterfall projects require a disciplined management as they are

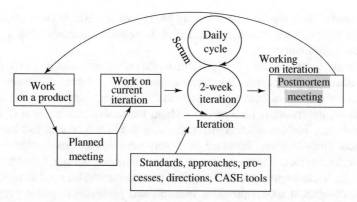

Fig. 10 Agile lifecycle

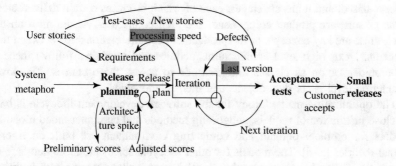

Fig. 11 Scrum lifecycle

Fig. 12 Extreme programming lifecycle

document-driven. In case of crisis, due to one-pass development of the software it is quite likely that the product does not meet the requirements of the customer.

Rapid prototyping, if used "as is", may tempt developers to reuse a quickly developed, untested, unreliable and undocumented prototype code as a product; this imposes extra risk constraints for any crisis implementation. However, prototyping potentially results in fast and economically efficient consensus on the customer requirements. Thus, we recommend using prototypes in crisis, in combination with the other models, such as spiral or waterfall, to promote maintainability and early return on investment.

The same considerations are applicable to the incremental model, as the product can gradually update to meet the requirements of the customer. However, due to evolutionary process of incremental product development, which requires an open architecture, the model is hard to use with an innovative product as it can easily degenerate into build-and-fix. Synchronize and stabilize model is risk-based and potentially crisis-adaptive; however, it is very sensitive to specific and complex testing technologies and tools. In case of crisis conditions, the spiral model is better suitable for in-house projects, as it requires specific knowledge on risk assessment. The object-oriented model provides iteration and parallelism; it also provides a better resource flexibility and thus is essential for crisis conditions. However, under a poor discipline the object-oriented projects are likely to degenerate into an expensive and unpredictable build-and-fix lifecycle.

6 Optimizing the Lifecycle: Enhanced Spiral Methodology

Every lifecycle stage of the software system development can be optimized, including requirement analysis, product specification, design, implementation, maintenance and retirement. To optimize the lifecycle, i.e. to adapt it for crisis conditions, a complex methodology is required. This section focus is the basic outline of the optimization methodology for the product lifecycle, which includes a set of models, methods, CASE tools and practices. The methodology is process-based, and it has six stages, each of which produces certain deliverables in terms of software product components and their connectors. At the most abstract level, these are key concepts of the product and certain relationships between these concepts. Next, high-level architectural modules and interfaces follow; these are elaborated later on as classes and methods to access these classes. The lowest abstraction level is for data objects and their relationships.

The optimization methodology for the software development lifecycle is based on close integration of models, supporting methods and computer-aided tools. The models for problem domain and computing environment are built on rigorous formal theories [2–6]. The models for other lifecycle stages are more heuristic and pragmatic. Therewith, the supporting development toolkit contains both traditional CASE tools and the so-called "lower" level tools, which integrate the formal model and the software product components.

The process diagram of the methodology for optimized software product development is to a certain extent similar to the spiral lifecycle model (Fig. 13). The methodology provides iterative bidirectional component-based development of open, expandable heterogeneous software products in global environment; it supports data consistency and integrity control. Heterogeneity involves architectural and structural aspects. The architectural heterogeneity means that the methodology

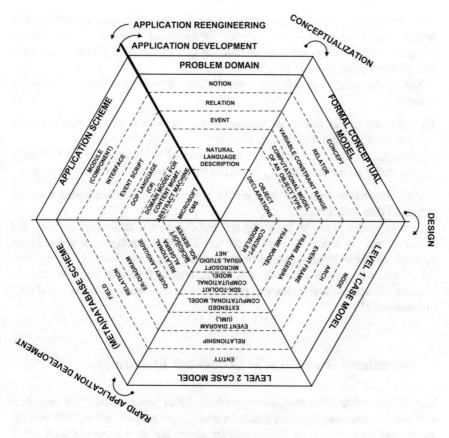

Fig. 13 Process diagram of the software development lifecycle

allows for integration of the modules or software subsystems, which are based on different architectures, such as mainframes, file servers, client servers and clouds. The structural heterogeneity means that the methodology allows for integration of the modules or software subsystems, which manage different kinds of data objects, such as relational databases, audio and video data and scanned documents.

During the software development lifecycle, the components of the heterogeneous software systems are transformed from problem domain concepts to formal model data entities. Further, by means of the software toolkit, which includes ConceptModeller [7] and content management system [8–10], the product is transformed into a complex semantic network and object-oriented warehouses managed by an abstract machine and represented by a virtual machine at the CASE level. Finally, we arrive to a well-formed layout of software product component interfaces managed by an internet portal superstructure. The development levels are elaborated in terms of entities, relationships, languages for content definition and management, and software tools.

A family of the formal object models for data representation and management supports the methodology for software development lifecycle. These models incorporate fundamental methods of finite sequences, variable domains, semantic networks and other theories [10–14].

The methodology for software development lifecycle provides the following features:

(i) Rigorous object models of heterogeneous software products, their elements and families, and their environments;

(ii) Integration of formal models, industry-standard technologies and CASE tools for software development by means of the innovative "middleware" tools.

Both advantages were implemented for representation and management of the integrated data and metadata.

Currently, the focus of mathematical and conceptual modeling, analysis and design of the software products shifts the lifecycle paradigm of the software development from object-oriented to pure object approach, i.e. from IT to computing. Computing is a relatively new research area; it models complex, heterogeneous, changeable and interactive problem domains in terms of objects and their environment [10].

7   Organizing the Lifecycle: Sequential Elaboration

The major purpose of the methodology is multi-factor optimization of the model for software development lifecycle, which, in crisis, is mission-critical for both product quality and project success. The key optimization factors for the software development lifecycle are: time, budget, requirements conformance and quality attributes, such as product performance, maintainability, security and the like.

Therewith, specific features of our understanding of the term "optimization" are the following ones. First, we do not mean optimization in common mathematically rigorous terms; instead, we select the best (or sometimes even a good enough) option out of a finite number of discrete values rather than a maximum of a continuous function. Second, the priority of the factors is dependent on the software project scale and scope. Third, the optimization factors are typically measurable and have certain metrics, such as number of code lines and defect removal rate. For each possible software solution, we can calculate the scenario-dependent optimization parameter values based on the above metrics and certain priorities. The resulting indicative values make the basis for better justified project management decisions, which include project plan estimates.

Naturally, in case of crisis, especially for mission-critical, large-scale, complex and heterogeneous products, it seems reasonable to use the above lifecycle methodology for data representation and management at the analysis and conceptual design stages.

We developed visual CASE tools to support the formal models for data representation and management and the processes of the product lifecycle phases, such as analysis, design, implementation, integration and maintenance. Specific workflow management tools based on document management system support the lifecycle processes for the product development. For each lifecycle phase of the product development, depending on the lifecycle model type and on the project scale and scope, these workflow management tools assist in generation and processing of certain document types, such as project plan, requirements checklist and unit test report.

Since the book is aimed at crisis management of the lifecycle processes for software product development, let us limit our scope to the overview of the methodology, i.e. models, metrics, methods, and tools, and focus on certain examples; the methodology itself is covered in more details in [9, 10, 15].

During the requirement analysis phase, optimization often results in generating requirements checklist, which is a simplified and less formal document, than the detailed product specification. However, irrespective of the type of the specification document, it should contain the lifecycle model chosen for the product development. The lifecycle model is a global parameter, which critically influences the product development plan.

The product development process is a sequential elaboration of the functional specification for the software product. In the above case, the product conceptual model is instantiated to obtain a more detailed product specification, which is elaborated later in the lifecycle. Further, we implement the architecture of the databases and other subsystems, which make the software product. In crisis, the software development lifecycle for mission-critical, large-scale, heterogeneous products is usually iterative, evolutionary and incremental, and every iteration provides further elaboration of the product functions (Fig. 13). In essence, the process outline is an improved spiral lifecycle model of software product development. However, in a number of cases, this process outline is elaborated depending on the product scale and scope or on the "project triangle" crisis optimization in terms of time, budget and functions. For instance, such a crisis optimization may result in the lifecycle reduced to a waterfall, where the software development is limited to a single pass through all of the lifecycle phases, or even to a build-and-fix model with incomplete lifecycle and simplified product documentation.

The further product lifecycle is optimized and elaborated in terms of system architecture, key technologies and development environment, which includes CASE tools and programming languages. The lifecycle optimization and elaboration process also addresses the existing software environment. The product developed should have certain quality attributes; for instance, it should be predictable, reliable, maintainable and, ideally, reusable.

In crisis, it is critically important to keep in mind that the lifecycle phase impact into the project economics is uneven. For example, the maintenance phase is the most expensive and challenging; it requires over 60 % of project time and budget [16].

Consequently, maintenance phase planning should be very accurate. However, coding contribution into the product lifecycle expenses is minimal. Consequently, coding planning should not usually take a long time. A well-justified combination of the software development methods and tools is essential for low-cost crisis development.

In certain cases, such as test termination after reaching a satisfactory error threshold, it is the project manager who makes the decision; however, the other cases, such as software retirement, usually require multi-side project evaluation.

Object-based approaches to software development often help to create interactive, distributed, open and expandable software products; they range from classical object-oriented to active objects and "pure" objects. As we know, according to the object-based approaches, the lifecycle phases of product development are flexible and have dynamically adaptive borderlines. However, even in crisis conditions, the object-based approaches require a disciplined management based on quantitative software engineering metrics and processes.

CASE tools help to validate the software in order to meet product specifications; they are often based on rigorous mathematical foundations, such as reliability statistical analysis and formal logics. Such CASE tools require a moderate level of mathematical training as they are typically designed for analysts and developers of a medium qualification level.

In crisis, essential preconditions of a product success include frequent functional prioritizing and sequential incremental elaboration.

Project specifications should be rigorous, logically correct and consistent, non-contradictory, complete in critical functional coverage, and transparently traceable.

For each lifecycle phase, software engineering requires rigorous and disciplined processes for the product development; clients and developers should strictly follow them. In crisis, developers should also follow development and documentation standards; otherwise, product development is at risk of becoming an unmanageable informal anarchy with an unpredictable result. That is why we suggest a methodology as a set of interrelated processes, methods and tools, which guides development of a requirement-matching, maintainable and high quality software even under such crisis challenges as changeable requirements, "on the fly" budget adjustments and other similar uncertainties.

The lifecycle optimization methodology is based on a thoroughly selected and tested set of models, software engineering methods and tools; it has been practically approved for developing large-scale, complex, heterogeneous and distributed software products.

The implementations of the methodology embraced a number of enterprises, such as ITERA International Group of Companies, including nearly 150 companies of over 20 countries and over 10,000 employees, the Institute of Control Problems of Russian Academy of Science, Russian Ministry for Industry and Energy, and a few others [17, 18].

8 Conclusion

In this chapter, we discussed certain lifecycle models of software development. These were build-and-fix, waterfall, incremental, object-oriented, spiral and a few others.

We also presented a more detailed description of the lifecycle models application to software development. We compared benefits and shortcomings of the models discussed. Some of the models that we discussed were one-pass and straightforward, others required a number of iterations. One key conclusion that we made was that there was no "silver bullet", i.e. a universal lifecycle model, equally applicable to any software product. That is why the lifecycle model choice was dependent upon product size and scope, and each project required a unique combination of features. In crisis, we recommended to combine prototyping with any of the other models discussed in order to achieve a common understanding of the key product features and to reduce project risks. The lifecycle model choice determined project economics, time to market, product quality and overall project success.

Another major takeaway we made is that the product success essentially depended on a number of human-related factors, which included vision of the critical product functions, transparent communication, feedback and a few others. We cover these human-related factors in more detail in Chap. 5.

We also analyzed applicability of the lifecycle models to large-scale, mission-critical software systems, especially in a crisis.

Finally, we introduced a methodology, which included a spiral-like lifecycle and a set of formal models and visual CASE tools for software product development. The methodology was designed to optimize the software product lifecycle. This is mission-critical in crisis; we cover the implementations in more details in Chap. 4.

The methodology was applied to large-scale, complex software products and to heterogeneous environments. In the next chapter, we present more details on the product development methodologies in terms of processes, roles and artifacts.

References

1. Boehm, B.: A Spiral model of software development and enhancement. IEEE Comput. **21**(5), 61–72 (1988)
2. Amemiya, M., Arikawa, S., Ishizuka, M., Ueno, H., Okuno, H., Kithashi, T., Koyama, T., Saeki, Y., Shimura, M., Shirai, Y., Tanaka, H., Tanaka, Y., Tamura, K., Tsujii, Y., Tsuji, S.: Knowledge Representation and its Use. Ohm Press (1987)
3. Brookshear, J.G. Computer science: An overview (10th ed.), Addison-Wesley, 2003
4. Rosen, K.: Discrete Mathematics and Its Application, 7th edn. McGraw-Hill (2011)
5. MackKay, D.J.C.: Information Theory, Inference, and Learning Algorithms. Cambridge University Press (2003)
6. Sipser, M.: Introduction to the Theory of Computation. Cengage Learning (2013)

7. Zykov, S.V.: ConceptModeller: a frame-based toolkit for modeling complex software applications. In: Baralt, J., Callaos, N., Chu, H.-W., Savoie, M.J., Zinn, C.D. (eds.) Proceedings of the International Multi-Conference on Complexity, Informatics and Cybernetics (IMCIC 2010), vol. I, pp. 468–473. Orlando, FL, USA. 6–9 Apr 2010
8. Nilsson, N.: Principles of Artificial Intelligence. Morgan Kaufmann, San Francisco (1980)
9. Sommerville, I.: Software Engineering, 9th edn. Addison-Wesley, 790 p. (2011)
10. Wolfengagen, V.E.: Applicative Computing. Its Quarks, Atoms and Molecules. Jurinfo-R, Moscow, 62 p. (2010)
11. Backus, J.: Can programming be liberated from the von Neumann style? A functional style and its algebra of programs. Commun. ACM **2**(8), 613–6412 (1978)
12. Minsky, M.: A framework for representing knowledge. The psychology of computer vision. In: Winston P.H. (ed.) McGraw-Hill (1975)
13. Barendregt, H.P.: The lambda calculus (rev. ed.), Studies in Logic, 103, North Holland, Amsterdam (1984)
14. Cheney E.W., Kincaid D.R.: Numerical Mathematics and Computing, 6th edn. Brooks/Cole (2007)
15. Ziegler, C.: Programming System Methodologies. Prentice Hall Inc, Englewood Cliffs, N. J. (1983)
16. Zykov, S.V.: Enterprise content management: bridging the academia and industry gap. In: Proceedings of the International Conference on Information Society (i-Society 2007), vol. I, pp. 145–152. Merrillville, Indiana, USA. 7–11 Oct 2007
17. Zykov, S.V.: The integrated methodology for enterprise content management. In: Proceedings of the International of the 13th International World Multi-Conference on Systemics, Cybernetics and Informatics (WMSCI 2009), pp. 259–264. Orlando, FL, USA. 10–13 July 2009
18. Zykov, S.V.: An integral approach to enterprise content management. In: Callaos, N., Lesso, W., Zinn, C.D., Zmazek, B. (eds.) Proceedings of the International 11th International World Multi-Conference on Systemics, Cybernetics and Informatics (WMSCI 2007), vol. I, pp. 212–216. Orlando, FL, USA. 8–11 July 2007

Chapter 3
Software Methodologies: Are Our Processes Crisis-Agile?

Abstract In this chapter, we discuss software development methodologies. These are adaptive process frameworks adjustable to software product size and scope. They usually include a set of methods, principles and techniques, and software development tools. Each of the methodologies can implement any of the lifecycle models. We discuss the difference between formal and agile methodologies. The formal methodologies include more artifacts; they have a rich set of complex processes, which include larger workflows and smaller activities. For each activity, every role assigned to it produces a deliverable. In crisis conditions, such as hardly formalizable problems, rapidly changing requirements and other uncertainties, agile methodologies, which are more adaptive and flexible in terms of artifacts, are applicable. The agile methodologies rely on self-disciplined and self-manageable teams, and consequently they are more constrained in terms of human-related factors. Similar to lifecycle models, there is no "silver bullet" in software development methodologies. Due to rigorous processes and more artifacts, formal methodologies are suitable for large-scale product development. Agile methodologies require special techniques and high level of discipline; otherwise, they can likely result in a low quality of software production.

Keywords Software development methodology · Formal methodology · Agile methodology

1 Introduction

The previous chapters introduced lifecycle models for software product development. In this chapter, we discuss a few software development methodologies. These are adaptive process frameworks adjustable to software product size and scope; they include a set of methods, best practices and tools.

Each of the methodologies that we will discuss can implement any of the lifecycle models introduced in the previous chapters. However, some of the methodologies are more formal in terms of development process and product artifacts,

© Springer International Publishing Switzerland 2016   51
S.V. Zykov, *Crisis Management for Software Development
and Knowledge Transfer*, Smart Innovation, Systems and Technologies 61,
DOI 10.1007/978-3-319-42966-3_3

while others are more flexible. In crisis conditions, which are usually more uncertain, agile methodologies are applicable. They usually require fewer product artifacts; however, disciplined team development is mission-critical.

Similar to lifecycle models, there is no "silver bullet" in software development methodologies. The formal methodologies are suitable for mission-critical and large-scale applications. The agile ones are often more crisis-resistant; however, in case of undisciplined development they may degrade into build-and-fix lifecycle and low quality software.

This chapter is organized as follows. Section 2 presents an overview of process frameworks of the software development methodologies. Section 3 presents the key features of Rational Unified Process methodology. Section 4 describes the Microsoft Solution Framework methodology. Section 5 provides an overview of the flexible methodologies, which are often mission-critical in crisis. The conclusion summarizes the results of the chapter.

2 Methodologies: Flexible Process Frameworks

The previous chapter described the lifecycle model of software systems, the main stages of their development, from a conceptual idea, through requirements specification, design, implementation, and maintenance or support to retirement. We discussed how these steps are performed in case of certain lifecycle models. Some of the models, such as object-oriented, include all of the lifecycle steps, while the others, such as build-and fix, do not. There are certain models based on linear sequence of phase changes, such as waterfall, while the others, such as object-oriented or spiral, support certain cyclic or iterative lifecycle processes. Some models, such as object-oriented, allow parallel or concurrent execution of some lifecycle stages, for instance, analysis and design. There are dependent lifecycle models such as rapid prototyping; therewith, it is often reasonable to combine rapid prototyping with the other lifecycle models, such as spiral or waterfall. Such combinations often help to significantly save the implementation time and costs, and to avoid severe design errors, as this helps to demonstrate the functionality of the software product in its early stages, such as analysis, preliminary design and requirements specification. Moreover, rapid prototyping helps to establish and maintain better interaction with the customer, since it makes easier to determine the mutually beneficial direction of product development and to verify that both parties understand the product features similarly. For the above reasons, the lifecycle model of rapid prototyping is especially valuable as a crisis management solution for software product development [1].

Let us consider methodologies. They are a parallel dimension to the lifecycle models; the methodologies assist in software product development of mission-critical systems [2–6]. Usually, a methodology means a set of techniques, models, methods and tools. Here, it will also mean a set of best practices for software development. Within a software development methodology, we can seldom find rigorous mathematical models other that these used for economic evaluation and feasibility study of the project. A number of approaches, especially in case of agile methodologies, such as Scrum or Agile, suggests a flexible set of customizable best practices, i.e. practical methods for software systems development. Therefore, it is often meaningless to consider a methodology as a purely theoretical research subject. In this respect, many of the above considerations on the lifecycle models for software systems development are applicable to the methodologies in case we include the best practices and certain classes of tools into the scope of our discussion.

In our view, a software development methodology is a parallel direction in relation to a lifecycle model. We have already mentioned the differences in lifecycle for software products and software projects; our focus is the products. A methodology is useful as a process framework and a set of practices for efficient software product (and project) development, including mission-critical applications and crisis conditions. Therewith, the methodologies that we are going to discuss can support various lifecycle models. For example, the Rational Unified Process (RUP) methodology can use either waterfall or spiral lifecycle as a basis. The other large-scale methodology that we are going to discuss, the Microsoft Solution Framework (MSF), also supports a number of lifecycle models. In terms of software systems development and the lifecycle stages, a methodology is a less formal approach than a model. It is often possible to scale a methodology up or down, since it is a framework, which depends on the size and scope of the software product. For example, Rational Unified Process initially intended for large-scale software development can scale, i.e. it can use more or less detailed development plan, processes and deliverables. Similarly, for the Microsoft Solution Framework there are more flexible implementations (often called MSF Agile) and more detailed ones (often called MSF Formal). An adequate size-to-deliverable ratio for methodologies and size-to-phase ratio for models is the key to crisis management of software product development.

According to the product size and scope, there are certain methodologies that are initially designed to build large-scale and mission-critical systems. We can call them large-scale, heavy, or formal. These are somewhat similar to the full-scale lifecycle models, which embrace the entire lifecycle and produce elaborate product documentation. However, each of these large-scale methodologies being a framework of principles, best practices, processes and deliverables, allows for a downsized implementation of software systems development. Additionally, there is a number of more flexible methodologies, which are suboptimal for large-scale and mission-critical software products, and which are designed to accommodate crisis conditions, such as high risks, requirements fluctuation and high uncertainty. The large-scale methodologies as process frameworks can support all the stages of the lifecycle models, such as the above-mentioned waterfall or any of the iterative patterns [7]. The large-scale methodologies are RUP and MSF. RUP is the standard of IBM, and MSF is the Microsoft standard. Interestingly, the MSF methodology originated from the synchronize and stabilize lifecycle model [8]. The MSF methodology is complex and agile at the same time; it supports multiple and highly scalable software development teams.

Potential benefits of MSF for crisis management result from a certain level of equality of the project team members, and clearly distinct personal responsibilities at the same time. Such team organization allows for scalability and crisis agility even in case of team of teams. In crisis, certain project team roles may overlap according to the trade-off matrix of roles. For example, project manager and product manager are typically different people; however, it is possible to combine their responsibilities for a small project or a local crisis. Likewise, some other role combinations are possible.

Common vision is another potentially promising concept of MSF, especially in crisis. Vision is an original idea, clear yet informal, of the fundamental differences and customer values for the future software product as compared to the existing ones, and its benefits after the implementation. A distorted vision can easily cause chaotic development and local crisis; however, sharing common vision through open communication is often a remedy for the crisis. The progress of software development is sequential elaboration of the software product, and, of course, the vision is the most abstract representation of it. Before the product development based on this representation actually starts, the developer side needs a high-level, and, later on, a more detailed document that describes the product specifications. A better common vision for the project team promotes more accurate product specifications. As soon as the project team is formed, the project schedule is developed. This includes roles and their activities for each product development stage. The project plan also includes the key activities duration, the primary control points, i.e. milestones, and the results at each stage of the product development, i.e. deliverables. After the planning phase, the development phase starts. In crisis, it is very important to detect the project activities and the product deliverables, which lie on the critical path in terms of resources including labor, timeframe and quality. It is equally important to ensure that the shared project deliverables will provide operational quality within the milestones set, at least for the mission-critical functions.

Often, a project team is recruited for the only project. Thereafter, it often happens that the product operates independently from the development team, as the maintenance team is different. That is why product documentation quality is critically important as it ensures continuous operation of the entire product line, such as Microsoft Windows, Office applications, and so on.

MSF as a methodology embraces not only lifecycle of a software system, but also the methods and techniques of software product and project development, including processes and roles in the project, responsibilities and deliverables, communication and teamwork, and project documentation.

Concerning agile methodologies, we are going to discuss Scrum, Agile and eXtreme Programming (XP). These are sets of best practices, i.e. recommendations for crisis management of software product development under high uncertainties and risks. Given the budget and timeframe, the aim of the agile methodologies is to develop the product of a certain quality level, or to cancel development if this is impossible. As in case of the lifecycle models, we will also describe the advantages and disadvantages of the software development methodologies.

Overall, the methodologies are practically oriented approaches focused on cost optimization. However, optimization in this context does not use a rigorous mathematical model, although the methodologies use certain product metrics to monitor, evaluate and plan the product development. Since there is no clear way to develop a mathematical model of software development with these methodologies, it is not entirely correct to say that they result in the optimal solution. However, with the help of processes and metrics the developers can reach sufficiently good and justified project decisions, which is still suboptimal in a mathematical sense.

3 Rational Process: Managing Mission-Critical Development

Let us discuss the Rational Unified Process methodology, or RUP. It was developed by Rational and inherited by IBM [9]. Currently, the product line of Rational CASE tools supports RUP; it covers all phases of the software product lifecycle. The product line includes over 10 tools for software design and project management, and a set of development and support tools for testing, implementation and maintenance. The tools interact with each other; they support RUP processes for the entire software product lifecycle.

In general, RUP is an iterative process methodology, which supports sequential elaboration of the product under such software development models as spiral or incremental. Risk assessment is an important component of RUP; in fact, this holds true for all above-mentioned methodologies, specifically agile ones. RUP is architecture-centric, i.e. the focus of the product development is the architecture and high-level design. Another key feature of RUP is use case diagrams; these are a part of UML standard supported by most CASE tools, they correspond to preliminary design stage of the software product lifecycle [10]. Further in the development lifecycle, use case scenarios instantiate the use case diagrams. For example, there are at least two scenarios even for a trivial use case, which include such roles as the user and the system, and represent a system login; these are a successful and an unsuccessful login attempts. Scenarios usually describe all possible instances of the use cases thus making software product predictable and usable, and helping to avoid local crises of unstable (and often undocumented) product behavior.

RUP and MSF contain a set of well-defined and structured software development processes. RUP development cycle includes four main stages; MSF generally includes four similar stages, plus stabilization. RUP and MSF are suitable for production of large-scale, mission-critical, component-based software systems. For scalability, RUP and MSF define multi-level processes of interaction between the project team roles; each low-level process stage has a deliverable, i.e. a software product artifact.

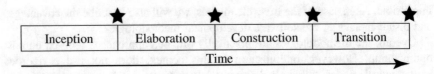

Fig. 1 RUP: stages

The four stages of RUP are inception, elaboration, construction, and transition (Fig. 1). Each stage can repeat several times (typically, at least three); these repetitions are called iterations.

The first stage of RUP is called inception; it includes the product concept, high-level system requirements and feasibility analysis in terms of time and cost. Of course, detailed design stage takes place later on; a requirements document at the inception stage describes the key functional requirements and constraints for the product. This requirements document at the inception stage also presents an overview of the project plan; this includes a list of the major project constraints in terms of timeframe and budget.

Milestone is an important concept of the inception stage. This is a key step required to complete the project; each milestone has a clear termination condition based on the deliverables, i.e. the documents to be produced before the milestone is reached. As soon as the deliverables, such as product concept, high-level product requirements and the draft project plan are complete, the inception phase is over, and elaboration stage starts.

Elaboration stage of RUP includes detailed design activities. Inception stage usually answers the question: "What is the product?" Elaboration stage typically addresses the question: "How do we build the product?" Elaboration stage delivers the architecture of the product; it describes which components will the product have, and how will they interact. From the perspective of software architecture, elaboration decides whether the product will include two or three tiers, what kind of a database will it use, and so on. In addition, detailed requirement specification is produced [11]. For example, in case of object-oriented lifecycle model, the deliverables will include complete list of all software product classes together with their signatures (i.e. names, types, and access modifiers for the attributes and methods), local and global variables, methods that will interact with the neighboring classes, and detailed algorithms and data structures [12].

As soon as the above deliverables are produced and detailed design is over, the third stage of RUP begins, which is construction. Construction stage includes product implementation, unit testing, integration and product testing. After the construction stage, the product is accepted and transferred to the customer. The final milestone of the construction stage is the product review, which ensures that the product meets customer requirements, and it is able to pass all acceptance tests at the client's site with the actual hardware, software and data. As soon as construction is over, transition stage starts.

Each of above stages may include a different number of iterations. Iterations, in their turn, are subdivided into activities, each of which is a relatively small isolated task with clear exit criteria. Each stage, such as construction or elaboration, can have several iterations; elaboration iteration will result in product redesign, and construction iteration will result in final readjustment before delivery [13]. Naturally, each iteration uses metrics and thresholds to control deliverables quality and to manage software production processes. For each iteration, there are work-flow processes, which include the key phases of software product lifecycle, such as analysis, requirements specification, design, testing, and so on. Similarly, each stage, such as inception, elaboration, construction and transition, may have several iterations. Each iteration, in its turn, usually includes a number of activities for several workflows.

We have already mentioned that the RUP methodology supports iterative and single-pass models, such as waterfall.

The waterfall lifecycle can be used with RUP in case of an approach somewhat similar to the incremental model. Suppose that a software product has a stable upgrade path and enables smooth functionality increase. For each RUP stage, we can use a waterfall-like approach. Each waterfall phase can use multiple iterations and prototyping. We may link the waterfall phase deliverables, such as conceptual, preliminary and detailed design documents, and the like, to each of the RUP stages.

Thus, a combination of waterfall lifecycle and RUP methodology is possible in case of iterative development application inside each of the waterfall phases and RUP stages. The waterfall lifecycle is a single pass, so it is no longer possible to make functional changes in the construction stage. The business and technological constraints are: predictable path of product development, clear definition of the functional requirements for the new release of the software product, and a sufficient project team experience in the problem domain and technologies of the product design and implementation [14].

An alternative RUP process may be based on the incremental lifecycle. This prescribes prototyping conceptual design at the first stage in order to verify the basic functionality. The second stage is the architectural design. The next stages, development and transfer, usually include a few iterations. With incremental life-cycle, it is required to define the number of releases, the number of iterations for each software product release, and the functionality for each release. To construct software product by gradually improving its functionality, there should be an open architecture and a predictable sequence of software upgrades from the previous releases to the next ones.

The RUP process framework includes so-called best practices; it can adapt to match a number of lifecycle models, such as waterfall, incremental, spiral, and evolutionary. However, a choice of a particular subset of these best practices often results in incorrect development framework, though the developers have the illusion that they use the methodology as prescribed. The best practices are designed to work together. One of the RUP best practices is iterative development. Complete product development in a single pass is usually not required. Product development happens in iterations; it includes requirement specification, architecture, detailed design and implementation, testing, integration, and transition. In each iteration, requirements are reviewed and adjusted. A general architectural requirement is a component-based architecture. This is essential for large-scale and mission-critical systems as they represent a set of interacting modules, each of which is designed to solve a relatively independent task. Component approach helps to design systems with minimal interaction of the components; this promotes interoperability and maintainability. Thus, any maintenance adjustment influences a single component or a small number of adjacent components. Even a large-scale system change usually requires a relatively small amount of labor and does not critically influence efficiency in terms of quality attributes; this is essential in crisis.

RUP usually requires visual CASE tools for modeling and design. Visual nature of tools is a requirement, because RUP uses UML notation with rigorous diagrams and graphic artifacts, such as classes, use cases and so on. These diagrams and graphic artifacts make the language for project management and product design in each iteration of every stage. In crisis, it is critically important to communicate in terms of the uniform UML notation supported by standard CASE tools; this provides a common interface for architecture design, code generation, system integration, unit testing, and other software development activities. Another essential requirement for RUP is continuous product quality and change management; this is also critical in crisis.

The RUP structure includes processes, roles and artifacts, which are the deliverables developed in each stage (Fig. 2). RUP uses a series of manuals, templates and instructions that guide the product development processes. For Rational CASE tools, there are design manuals, which include the templates for the product artifacts development.

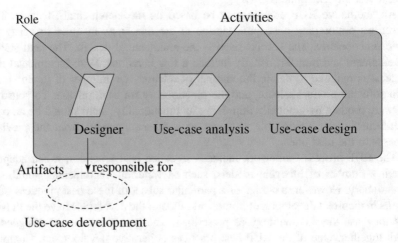

Fig. 2 RUP structure: roles, activities, artifacts

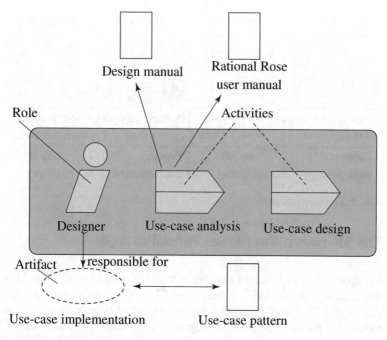

Fig. 3 RUP structure: manuals, patterns, manuals

RUP also describes the working processes and their details in terms of activity diagrams and process manuals (Fig. 3). RUP focuses on continuous risk monitoring and feedback, critical risks management, requirements conformance, change management, architecture-based development, component design, teamwork and quality management.

RUP can adapt to sequential and iterative lifecycle models, such as waterfall, evolutionary, incremental and spiral. RUP is scalable in terms of artifacts, activities, and roles; it can be more or less formal (Fig. 4). Therefore, we can say that RUP has certain crisis agility, and that RUP becomes "light" or "heavy"/formal depending on product size. Perhaps, RUP is less applicable for very small products, because it often requires significant labor costs and expensive CASE tools. In crisis, RUP may become inefficient for small-scale products: it generally requires overhead costs associated with the detailed product documentation, staff training and risk assessment. However, RUP as a formal approach is critically important for medium and larger products, which usually exceed 100 KLOC; its artifacts are well-defined, and the processes are formal and document-driven. For crisis agility, we recommend a less formal RUP and prototyping.

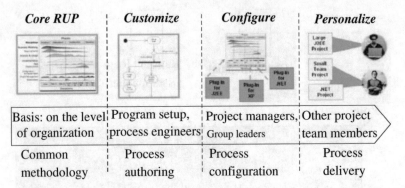

Core RUP	Customize	Configure	Personalize
Basis: on the level of organization	Program setup, process engineers	Project managers, Group leaders	Other project team members
Common methodology	Process authoring	Process configuration	Process delivery

Fig. 4 RUP: Process adjustment

4 The Microsoft Way: From Formal to Agile

Another approach, which is also scalable in terms of product size, is the Microsoft Solution Framework, or MSF; it originated from the synchronize and stabilize lifecycle model. However, MSF also includes a set of best practices, i.e. recommendations for software product development, according to a certain set of roles and stages. Similarly to RUP, MSF is a framework, i.e. it includes a customizable and flexible set of multi-level processes and recommendations for software product development. Currently, MSF supports a number of lifecycle models rather than the only synchronize and stabilize model.

The basis for MSF is a large and versatile 20-year experience of the Microsoft Corporation in software product development, which includes global distribution, millions of customers, and such products as Windows operating system and Office applications. Importantly for us, the basis for MSF is not only successful projects, but also a number of challenging ones. MSF uses a set of Microsoft products for project management and software development, which includes such CASE tools as Visual Studio, Project Server and SharePoint. MSF supports projects of different scale; it is flexible and adaptive. Therewith, the approach is extremely interesting for crisis management of software product development.

RUP and MSF are process-based frameworks. Similar to RUP, MSF supports not only large-scale but also smaller scale products; it has two distinct process frameworks, Formal and Agile, for the former and latter ones. The other similarity to RUP is that MSF supports a number of visual CASE tools for design, implementation, testing and maintenance, i.e. for the entire software product lifecycle. Microsoft Operation Framework is an add-on to MSF; it supports software product maintenance.

The key MSF CASE tool is Visual Studio; it supports visual development of components or assemblies written in different programming languages. Visual Studio provides a universal storage for project metadata; it facilitates team development, service-based software production and secure component design, where unique digital signatures guard each of the assemblies.

Let us consider the basic elements of MSF; they are similar to that of RUP, however, there are some notable distinctions. MSF suggests a more team-focused methodology than RUP; along with the development process framework, it supports a scalable approach for software product teamwork. Microsoft has a vast experience of software project management, including a number of development and support centers, and the largest knowledge base for beta testers and end users assistance.

MSF includes a number of documents that define the processes for project management, including policies for the key roles, such as project manager and product manager, risk management plans and mitigation procedures using the contradiction management matrix, to name a few. Additionally, MSF includes a set of practical recommendations for the two particular instances of its general framework, Agile and Formal. Each of the two instances, Agile and Formal, has its own specific features; however, they both focus on team development, clear separation of concerns in roles, and transparent communication. In contrast to RUP and some other methodologies, MSF assumes a relative equality of the project roles in terms of communication. For instance, a junior developer's opinion is always considered along with the project manager's one. Opinions of all project team members are taken into account.

The general framework of MSF process workflows and activities, which are smaller process steps of the workflows, is somewhat similar to RUP. However, MSF features specific reports related to the deliverables for the work items, i.e. the tasks within an activity. Each task has a number of states, and the task report usually features a checklist, which indicates the degree of completion of the task. The Formal instance of MSF includes a larger number of artifacts, documents, and a larger set of roles; it is more suitable for development of large-scale and mission-critical software products.

Table 1 shows the relationships between the key MSF elements and the principles that underlie the methodology. MSF is process-oriented and risk-driven. Detection and monitoring of key risk factors are important practices. MSF recommends building databases that store risk assessment results. Another important item is the knowledge accumulated in previous projects. The postmortem reports written after each stage of the process are stored in the forecast database for the future projects.

The best practices of MSF instantiate the key ideas of teamwork, knowledge acquisition and analysis, risk assessment, and flexibility. These best practices include partnership with the client and transparent communication. Transparent communication is critically important for a crisis project: poor communication often results in misinterpreting or misunderstanding of the information, which may lead to a local crisis. The project team needs to communicate in order to share the

Table 1 MSF elements and their relations

Basic principle	Model or discipline	Key concept	Pre-tested practice	Recommendations
Learn by any experience	Process model	Desire to learn	Post-phase review	External trainer participates
Take care—wait for changes	Risk management	Continuous risk evaluation	Detecting and monitoring risk factors	Creating database on risks

conceptual framework, to establish common vision and to understand the ultimate goal of the software product. Other best practices suggest maintaining quality, flexibility, proactive adaptability to changes in functional requirements and constraints, and creating value. One of the MSF key concepts is a deliverable, i.e. any measurable project artifact, which is a result of any project task, work item or activity.

MSF team has a number of key roles, which include clearly separated responsibilities. However, some roles may overlap in terms of certain individuals participation in the project. For instance, MSF allows combining the roles of project manager and product manager. Certain combinations are possible (though not recommended), while the others are not. MSF usually represents the recommendations for combining roles in the form of a matrix, which is often helpful for team optimization in crisis (see Fig. 6).

Figure 5 represents MSF team in terms of disciplines or knowledge areas. These include program management, product management, user experience, architectural design and maintenance.

The MSF project team is the team of equal, that is, everyone is equally welcome to contribute to project success. However, transparent communication and clearly separated responsibilities for each role are also critically important. When it comes to the quality of the product as a whole, MSF encourages open communication, and each team member is equally responsible for his or her part of the result. Depending on the deliverables, each role has clear metrics of quality control. In fact, the progress of the project depends on a well-balanced analysis of the contribution of each team member. In order to ensure accurate, balanced product development, to prevent and compensate for the crisis consequences arising from errors, inconsistencies or uncertainties, the team has to take into account every significant aspect of the future product. Therefore, each team member has an equal right to vote and contribute. For crisis product development, a very important MSF principle is agility, which means that the process framework is scalable in terms of the number of team members, their functional roles, constraints, areas of knowledge and deliverables. In certain crisis cases, a project team can be subdivided into a number of smaller teams; this makes MSF dynamically resizable, adaptive and more efficient in terms of communication.

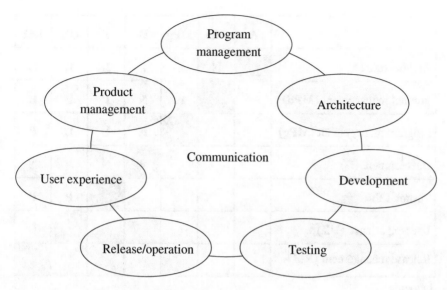

Fig. 5 MSF team: knowledge areas

Role is a central concept in MSF. Each role carries out certain activities that make action sequences, or workflows. An activity may include several tasks; these are even smaller project fragments with clear deadlines and measurable deliverables. An activity may result in a (partial) product, and it may require another partial product as an input. Certain activities can be concurrent; this is similar to product development phases of the object-oriented lifecycle model. A typical deliverable for an activity is a document, such as a table, a graph, a diagram, a text of a manual and so on. The main knowledge areas, which form the project roles, are architecture, product management, development, testing and maintenance.

Figure 6 shows the matrix of possible roles combinations, which illustrates MSF project team scalability. For instance, it is possible to combine certain roles, such as product manager and project manager; however, the matrix does not recommend this combination. Thus, an MSF project team can combine certain roles; this helps to optimize the team size in case of crisis.

Like RUP, MSF is process-based, and the phases of the two methodologies are similar. For instance, inception stage of RUP corresponds to vision stage of MSF, since they both contain the key idea of the product scope. The process models of RUP and MSF support iterative elaboration of the product and construction of several sequential releases. Each release produces an operational product; however, it may be functionally incomplete.

The exit criteria for MSF vision stage is setting of the scope, i.e. conceptual design constraints that delimit the core functionality of the software product. The scope includes the problem that the product should solve and its functional limits; it also defines key functions for further releases. The next stages are project planning and product development; they correspond to elaboration and construction in

	A	MD	MPg	D	T	UX	RM
Architecture (**A**)		N	P	P	L	L	L
Product management (**MPd**)			N	N	P	P	L
Program management (**MPg**)				N	L	L	P
Development (**D**)					N	N	N
Testing (**T**)						P	P
User experience (**UX**)							L
Release management (**RM**)							

Legend:

P – Possible

L – Low probability

N – Not recommended

Fig. 6 Matrix of MSF role compatibility

RUP. Each stage has a milestone, where the exit criteria are checked, and the results are compared against the estimated constraints and documented requirements. Once the planning is complete, development starts. The stage that follows development is specific for MSF, and this is stabilization (Fig. 7). Since MSF basis is the synchronize and stabilize lifecycle model, release stabilization is an essential part of the methodology. To verify that the release operates in a sustainable and reliable manner, it is tested as prescribed by the process and metrics.

After MSF release stabilization, the transfer occurs; this corresponds to RUP transition process. Upon MSF product transfer, the deployment stage begins. In case of poor discipline, which can be a source of a local crisis, MSF stabilization stage may result in a significant waste of time and labor. MSF stabilization includes a number of complex processes, so it is seldom used outside Microsoft. Generally, MSF is similar to RUP, as it is process-based, iterative, and has similar stages.

Under MSF, the project timeline is divided into separate stages with clear milestones and deliverables. The exit criteria for each milestone is based on product development metrics. After each release of the iterative development cycle, the product functionality is elaborated. The MSF best practices are proven techniques, which were developed and approved in a large number of previous projects. To scale

Fig. 7 MSF process model

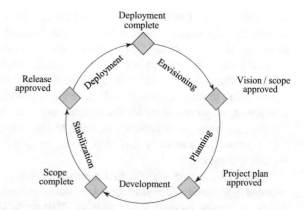

up for a large product development, relatively small teams are combined into bigger ones. Every function of a release is assigned to a certain part of the team; however, responsibility for each deliverable is strictly personal. The team members share open communication and equal rights. To meet a changeable product scope, MSF allows combining certain roles. The contradiction management matrix adaptively defines optimization priorities based on risks, resources and functionality.

5 Flexible Methodologies: Adding Crisis Agility

In crisis conditions, more responsive and adaptive techniques and practices than formal RUP and MSF are sometimes required. These are called agile; they include Scrum, Extreme Programming (XP), specifically Agile, and a few others. Agile methodologies require high level of software development discipline and a number of "soft" skills, such as teamwork, communications, negotiations, to name a few. The clear trade-off, however, is that the developers often have to sacrifice certain software product artifacts and a number of deliverables in favor of agility, i.e. crisis adaptability. Agile methodologies are risk-focused and iterative; they typically require a few incremental releases, each of which contains several brisk iterations. The agile documentation artifacts are often simplified to product backlog (i.e. prioritized list of activities) and user stories, each of which is a small paper card with a one-sentence description of a high-level functional requirement. Due to lack of resources, agile development teams often use oral communication. The client is usually an active participant of any agile development process. In agile, the client sets the critical functional requirements to be implemented first, and monitors product quality. The agile process facilitator is not a project manager; he or she has no administrative power to control the development. An agile team is self-manageable; this requires a disciplined product development, a high level of self-confidence and strict following commitments [15]. Besides user stories, agile project artifacts usually include backlog and "earned value" progress diagrams; there are simple standards for product documentation. The features of agile methodologies are: transparent communication; simple practices, standards and artifacts; adaptive self-managed team that cooperatively creates value; daily teambuilding through meetings, and active client participation [16]. The planning is transparent, no problems are hidden in agile, and every team member assists in their solution; no overtime is allowed. In agile, testing often precedes coding to ensure defect prevention and early detection. Agile methodologies prescribe prototyping for risk management.

Historically, Scrum was the first agile methodology to appear [16]. Scrum was intended for hard-to-formalize problem domains and rapidly changing requirements. A formal methodology, such as RUP or MSF, would require a certain amount of time and effort to produce the deliverables and to follow the processes; however, in crisis the requirements were likely to change long before the product delivery. Adaptive teamwork is a key success factor of any agile methodology. Since Scrum originated from movie industry, certain terms for processes, milestones and deliverables were also inherited. The Scrum development stages are: pre-play, play (including product review and correction) and post-play; they correspond to planning/design, implementation and postmortem/wrap-up before the product transfer. "Scene shot" means release done. The live "backlog" contains prioritized activities for the current "sprint", or iteration, which usually takes one week only. A short daily meeting helps in risk assessment and backlog prioritization, so that mission-critical functions come first. The name for the process facilitator is Scrum master; he or she is usually the most experienced team member in terms of specific techniques and processes.

Another agile methodology, XP, started from automobile production management and was later applied to software development by Beck [17]. The values of XP are: communication, simplicity, feedback, courage and respect. The first two values are similar to Scrum open communication and simple artifacts. The idea of open communication is to establish and disseminate common vision among the developers. To achieve the common vision, XP recommends metaphors, which are words or phrases used to provide a clearer description of an initially ambiguous idea.

Simplicity recommends to improve design and implementation by means of refactoring, which is gradual artifact quality improvement while maintaining functional correctness and behavior. Feedback sources are: product, team and customer; all of them help developers to estimate progress and adapt to changes. However, feedback should be done on time. Courage means self-confidence of the artifact author in reviews; it helps refactoring or even redeveloping the artifact from scratch. Respect means striving for product quality based on continuous refactoring; developers must provide only good quality artifacts to the team.

Pair programming and common code possession are among the XP practices; they are potentially efficient for sharing product vision and mitigating the risks of team member loss. Another interesting practice is that coders estimate their own labor. Continuous integration and quick release production are also key XP practices; developers must use all practices together, otherwise XP often fails. Product development includes larger releases, each of which consists of smaller iterations.

Common room for team development helps to track the project progress; wall diagrams of the backlog and earned value are shared. Product transfer is potentially smoother, since the client usually develops acceptance tests.

Among XP challenges are: dependence on oral communication, client's obligatory team membership, interdependence of the practices (some of which, such as pair programming, are human factor-dependent), and compact design standards in terms of the required artifacts, including product architecture.

A more recent flexible methodology is Agile; it started in 2001. The Agile Manifesto prescribes domination of people, communication, code, client cooperation and adaptive changes over processes, tools, documentation, contracts and planning [18]. Agile is a two-level, iterative, test-driven, empirical, feedback-oriented and adaptive team-based methodology.

The agile methodologies also include Crystal Clear, Adaptive Software Development, Feature-Driven Development and a few others.

The agile methodologies may become beneficial in crisis; however, in case of undisciplined development they may easily degrade into low quality deliverables, non-responsible team behavior and build-and-fix lifecycle.

6   Conclusion

In this chapter, we discussed a few software development methodologies. These are adaptive process frameworks, adjustable to software product size and scope. Usually, they include a set of methods, principles and techniques (also known as best practices), and software development tools.

Each of the methodologies discussed can implement any of the lifecycle models introduced in the previous chapter. Methodologies that are more formal include Rational Unified Process (RUP) and Microsoft Solution Framework (MSF). In crisis conditions, such as hard-to-formalize problem domain, rapidly changing requirements and other uncertainties, agile methodologies are also applicable; these include Scrum, Extreme Programming, specifically Agile and a few others. The agile methodologies have fewer artifacts and require self-disciplined development team; this imposes extra human factor-related constraints.

Similar to lifecycle models, there is no "silver bullet" in software development methodologies. Due to rigorous processes and well-defined deliverables, MSF and RUP are suitable for mission-critical and large-scale applications. The agile methodologies require disciplined development; otherwise, they may easily degrade into low quality software product and build-and-fix lifecycle.

References

1. Robertson, J.: Essential intranets: inspiring sites that deliver business value, 2013, Step Two Designs, 280 pp
2. Tanenbaum, A.S, Steen, M.V.: Distributed systems: principles and paradigms, 705 pp, 2nd edn. Pearson (2007)
3. Booch, G., Maksimchuk, R.A., Engel, M.W., Young, B.J., Conallen, J., Houston, K.A.: Object-Oriented Analysis and Design with Applications, p. 720. Addison-Wesley, New York (2007)
4. Lipaev, V.V.: Software Engineering. Methodological Foundations, 680 pp. TEIS 2006 (in Russian)
5. Pressman, R.S., Maxim, B.R.: Software Engineering: A Practitioner's Approach. McGraw-Hill (2015)
6. Sommerville, I.: Software Engineering, 790 pp., 9th edn. Addison-Wesley (2011)
7. Fowler, M.: Patterns of Enterprise Application Architecture. Addison-Wesley (2003)
8. Schach, S.R.: Object-Oriented and Classical Software, 688 pp., 8th edn. McGraw-Hill (2011)
9. Kruchten, P.: The Rational Unified Process: An Introduction, 336 pp., 3rd edn. Addison Wesley (2003)
10. UML standard: Retrieved November 25, 2015 from www.omg.org/spec/UML/
11. Elmasri, R., Navathe, S.B.: Fundamentals of database systems, 1201 pp. Addison Wesley (2011)
12. Kirsten, W., Ihringer, M., Röhrig, B., Schulte, P.: Object-Oriented Application Development Using the Caché Postrelational Database. Springer (2001)
13. Boucadair, M., Jacquenet, C.: Handbook of Research on Redesigning the Future of Internet Architectures, 621 pp. IGI Global (2015)
14. Coats, R.B., Vlaeminke, I.: Man-Computer Interfaces: An Introduction to Software Design and Implementation. Blackwell Scientific Publications (1987)
15. Leffingwell, D.: Agile Software Requirements: Lean Requirements Practices for Teams, Programs, and the Enterprise. Addison-Wesley (2011)
16. Nonaka, I., Takeuchi, H.: The knowledge-Creating Company. Oxford University Press, New York (1995)
17. Beck, K.: Extreme Programming Explained: Embrace Change, 2011 pp. Addison-Wesley (1999)
18. Manifesto, A.: Retrieved November 25, 2015 from http://www.agilemanifesto.org (2001)

Chapter 4
Software Patterns: Ready for Crisis Development?

Abstract In crisis, resource efficient software production is mission-critical; it includes management of requirement changes and release updates. We suggest a methodology of pattern-based software product development, which includes a set of formal models, processes, methods and tools. The methodology uses resource efficient component development based on high-level architecture patterns with certain combinations of baselines and branches. We support these development patterns by domain-specific languages and visual tools. Another challenge and possible source of crisis is development of large-scale distributed heterogeneous applications; in this case, we suggest an incremental software development methodology, which includes a set of models, methods and supporting tools. The methodology proved to be particularly efficient in terms of time, budget and quality for large-scale heterogeneous products. The areas of implementation included oil-and-gas production, air transportation, retail network and nuclear power generation. Each implementation used a domain-specific language to facilitate pattern-based product cloning, maintenance and re-engineering.

Keywords Domain-specific language · Architecture pattern · Heterogeneous application · Pattern-based development

1 Introduction

In crisis, resource efficient software production is mission-critical. In this chapter, we will discuss a methodology of pattern-based software product development, which includes a set of formal models, processes, methods and tools.

For resource efficient development, we will use high-level architecture patterns for frequently used large-scale product modules or their combinations. We will discuss how such patterns combine baselines and branches to manage such crisis uncertainty as frequent requirement changes. We will illustrate how such patterns use domain-specific languages and visual software development tools for better usability.

© Springer International Publishing Switzerland 2016
S.V. Zykov, *Crisis Management for Software Development
and Knowledge Transfer*, Smart Innovation, Systems and Technologies 61,
DOI 10.1007/978-3-319-42966-3_4

We will consider large-scale distributed heterogeneous applications and suggest a uniform incremental software development methodology, which includes models, methods and tools. We will discuss the methodology implementations for integration of different types of heterogeneous enterprise components, including legacy systems and weak-structured data.

The implementations will use domain-specific languages, which provide significant crisis agility and substantial savings in terms and costs for series of similar implementations.

This chapter is organized as follows. Section 2 presents an overview of high-level software development patterns for mission-critical systems. Section 3 introduces the key features of the new methodology for data lifecycle management. Section 4 describes an improved software development framework by means of adding design patterns. Section 5 provides a set of large-scale software product implementations based on patterns. These include an enterprise content management system for an oil-and-gas group, a domain-driven messaging system for a retail network, an air traffic control system and a computer-aided design system for a nuclear power plant. The conclusion summarizes the results of the chapter.

2   High-Level Patterns: Designing Mission-Critical Systems

Development of large-scale and mission-critical software products is a serious challenge, especially in crisis. The avalanche of bulky data hampers the development processes; the data is heterogeneous in structure and architecture. To conquer complexity and heterogeneity, a methodology is required, which includes rigorous mathematical models, development technologies and CASE tools. This methodology should embrace the entire product development lifecycle. The methodology should simplify product maintenance due to component-level metadata pattern management, which allows intensive artifacts reuse and which easily adapts to changeable business requirements. This section presents the methodology outline, which is based on pattern management.

An essential benefit of the approach is adaptive, sequential elaboration of the heterogeneous component interaction schemes for the product in order to match the rapidly changing business requirements. Such benefit is possible due to the reverse engineering feature of the process framework of product development lifecycle (Fig. 13 in chap. 2). The reverse engineering is possible down to the formal model level [1]; it allows rigorous component-based (and even object-based) software product verification [2]. Therewith, the methodology adds "optimized" flexibility to software product re-engineering and verification, in accordance with the business requirements specified. This is possible because of rigorous modeling of the iterative, evolutionary process framework for software product development.

The process model framework for software product development also allows building a catalogue of templates for heterogeneous software products based on an integrated metadata repository, which stores data snapshots. Thus, the process framework provides a solution to store relatively stable or frequently used configurations of heterogeneous software products, or certain fragments for such configurations. The approach potentially allows substantial decrease of the integration expenses for commonly used software product components and/or combinations, such as system modules of SAP R/3 and Oracle e-Business Suite ERP applications. In case the developer's metadata repository already contains a solution similar to the system the client requires, the approach allows essential savings in terms of budget, time and labor.

The above consideration clears the way for metadata repository development, which stores the chronological sequence of software product configuration snapshots as a tree with the baseline version and the branches for software product variations. This is somewhat similar to the functions of software engineering tools for version control [3, 4] . For each of the lifecycle phases of the product development, such as design, implementation and maintenance, the approach allows for a justified, "optimized" selection of the most adequate deliverables and for similar solutions based on the repository data clones.

For example, let us consider a mission-critical enterprise software product. In this case, the developer may generate the "clones" for different clients with similar requirements on heterogeneous ERP module combinations, or for different companies of a single enterprise. A visual illustration of the latter solution is a tree of the company portals based on shared enterprise data warehouse, which stores the heterogeneous components of the commonly used ERP and CRM modules.

Let us discuss the prospective areas of the repository development for the metadata snapshots. To describe the metadata warehouses according to the product requirement specifications, it seems reasonable to develop problem-oriented meta-languages of domain-specific (DSL) type [5]. Two of these DSLs should specify the meta-warehouse structure and requirement specifications. We cover the formal models in more detail in [6–9]; they provide cross-referencing for the concepts of these two domain-specific meta-languages, and promise a more rigorous mapping in future.

Semantic network-based search mechanisms with frame visualization are likely to assist in building metadata snapshot repository of prospective software products; this provides a better matching for the future requirement specifications. The approach potentially allows time-and-cost-effective transforming of the existing software product components in order to match the new requirements with minimum corrections and, consequently, minimum labor expenses.

In case of a rich metadata snapshot repository, it is possible to reuse typical software product components or their common combinations for the current client or for prospective customers. Selection criteria for such common components could be percentage of reuse, ease of maintenance, client satisfaction, degree of matching business requirements, and the like.

Implementation of the suggested approach resulted in a unified software architecture, which integrated a number of heterogeneous components: state-of-the-art
Oracle-based ERP modules for financial planning and management, a legacy
human resource management system and a weak-structured multimedia archive.
The architectural framework of internet and intranet portals, which managed
heterogeneous content of enterprise-level software, provided a number of successful
implementations in ITERA International Group of companies, which had around
10,000 employees in nearly 150 companies of over 20 countries.

The methodology for pattern-based development of enterprise-level software
product frameworks includes formal models, software engineering tools and portal
architecture; it provides integration with a wide range of state-of-the-art CASE tools,
such as IBM Rational, Microsoft Visual Studio .Net and Oracle Developer [10–13].

Strategic benefits of the approach as compared to methodologies of the
above-mentioned vendors are possible due to integration of formal model and
software engineering tools, which are aimed at heterogeneous high-level
pattern-based metadata warehousing.

The data management challenge resulted from architectural and structural data
heterogeneity, and complexity of the integrated software solutions. The qualitative
assessments of the approach were based on major macro level indices: total cost of
ownership, return on investment and implementation costs. In terms of the
above-mentioned indices, implementation results in ITERA Group outperformed
the commercially available solutions by the average of 30–40 %.

Other implementations of the methodology included governmental and commercial enterprises, such as Russian Ministry for Industry and Energy, Institute of
Control Systems of Russian Academy of Science, and a few more [8, 9].

3   Data Lifecycle Management: Customizing the Process Framework

To illustrate how the methodology was implemented, let us focus on large-scale
product implementations and summarize their specific features.

Currently, the multinational enterprises possess large and geographically distributed infrastructures, which are typically focused on common business goals.
Each of these enterprises has accumulated a tremendous and rapidly increasing data
burden; the data growth rate is comparable to an avalanche. For certain enterprises,
their data bulk already exceeds petabyte size, and it tends to double every five
years.

Undoubtedly, management of such data is a dramatic challenge. The problem
becomes even more complex due to heterogeneity of the data, which varies from
well-structured relational databases to non-normalized trees and lists, and to
weak-structured multimedia data. Therewith, enterprises are facing crisis of efficient
data management and maintenance. Let us discuss how the methodology for

software product development assists in more efficient and uniform management of heterogeneous enterprise data.

The methodology includes a set of mathematical models, methods and the supporting software engineering tools for object-based representation and manipulation of heterogeneous enterprise data.

Direct application of the software development methodologies, such as RUP or MSF, to heterogeneous enterprise data management, results either in unreasonably narrow single-vendor solutions, or in inadequate time-and-cost expenses. Clearly, these methodologies would significantly benefit from a formal model. Similarly, purely formal approaches to data integration and management, such as Cyc and SYNTHESIS, do not result in acceptable quality implementations in terms of scalability, robustness, usability etc., since they lack CASE level tools support [14–19].

Thus, we suggest a methodology for development and maintenance of heterogeneous enterprise software products. The approach is based on rigorous formal models; it is supported by software engineering tools that provide integration to standard enterprise CASE tools, which commonly support such software development methodologies as RUP and MSF. The methodology eliminates data duplication and contradiction between the integrated modules, thus increasing the robustness of the enterprise software systems. The methodology integrates a number of levels for enterprise software systems development: data models, software applications, development methodologies, CASE tools and databases. The methodology includes process framework, a set of object models and CASE tools for data representation and management [9, 20]. The process framework for enterprise software systems development provides iterative bidirectional construction with re-engineering. The object nature of the "class-object-value" model framework provides compatibility with traditional object-oriented analysis and design, and with a few other approaches, including D.S. Scott's variable domains [21]. The formal model allows data representation of the enterprise software systems for internet environments.

The methodology relates product development stages to component-based data representation and component interfaces in the global environment.

The process framework sequentially transforms the data in the following order: finite sequence objects in the form of lambda calculus terms; higher order logical predicates; frames for visual data analysis; class definitions in terms of XML object schemes generated by the ConceptModeller CASE tool; data scheme as a set of UML diagrams [7, 9, 20, 22–24]. The above data representations are stored in the enterprise software content database, which includes heterogeneous data and metadata objects.

The enterprise content representation is based on semantic network situation model, which provides intuitive transparency for problem domain analysts when they construct the problem domain scheme. A frame-based notation visualizes the model objects.

The content management model is a state-based abstract machine with role assignments, which generalize the process framework for similar portal engineering activities, such as template generation, publication cycle, and role-based data access management. The abstract machine language contains functions, which model the major content management operations, such as declaration, evaluation and personalization. The language has a formal syntax and denotational semantics in terms of variable domains.

The architecture of the heterogeneous enterprise content management system provides unification due to uniform data representation of object association-based relationships at the data and metadata levels. The uniform enterprise content management for heterogeneous objects is based on the portal, which serves a meta-level superstructure over the enterprise data warehouse. Assignments act as code scripts; they change states of the content management abstract machine and provide dynamic, scenario-driven content management.

According to the process framework, the ConceptModeller visual CASE tool assists in semantically-oriented development of the integrated data scheme for the heterogeneous enterprise software systems [20]. The ConceptModeller tool uses a semantic network-based model; the model works in nearly natural-language terms, and it is intuitively transparent to problem domain analysts. Frame notation is used for visual representation of the data scheme for the enterprise software systems [25].

Deep integration with mathematical models and enterprise CASE tools provides an iterative, incremental software development lifecycle with re-engineering. The abstract machine-based tool is used for problem-oriented visual content management in heterogeneous enterprise software systems. The content management tool features a flexible scenario-oriented content lifecycle and role-based mechanisms. The content management tool provides a uniform portal representation of heterogeneous data and metadata, flexible content processing by various user groups, high security, ergonomics and intuitively transparent complex data object management.

4 Adding Patterns: Further Improvement of the Process Framework

The process framework for software product development [7] potentially allows for the following benefits. The framework includes a lifecycle, which is similar to spiral development model. The framework supports product verification, including requirement tracing. The framework supports a repository of metadata snapshots for enterprise software systems; the repository supports system rollback to virtually any previous state of the components architecture. The framework supports catalogue of patterns for heterogeneous software products based on the repository of metadata snapshots, which includes the snapshot states [8]. The framework supports cloning of the heterogeneous software products based on the repository of metadata, which contains the basic configuration and the "branches" for slightly varying enterprise software products. The framework supports a formal language specification to model specific domains [7]. The framework supports component-based adjustment of the existing enterprise software systems by using the metadata snapshot repository in order to match requirement changes. The framework supports reuse of the required components. Due to flexibility and agility, the above features make the framework crisis-adaptive.

The process framework supports waterfall; however, in crisis, we recommend incremental software product development with prototyping.

Bidirectional organization is essential for the process framework of the enterprise software systems development. The approach provides component-level reverse engineering possibility for the enterprise software. The formal model allows uniform function-based verification of heterogeneous components of the enterprise software products. The model is practically independent upon the hardware and software environment of the particular component [7]. The formal models represent system entities for the enterprise software in terms of "pure" objects. This is relevant to .Net and Java technologies, where any program entity is also an object.

The approach allows incremental adaptation of the component management schemes for the heterogeneous enterprise software products in order to match the rapidly changing business requirements. Such benefit is mission-critical in crisis; this is possible due to the reverse engineering feature of the process framework for enterprise software development. The approach supports reverse engineering down to model level, which allows rigorous component-level software product verification. In crisis, the approach enhances traditional re-engineering and verification by means of flexible correction and "optimization" of the enterprise software systems in accordance with the specified business requirements.

Another benefit of the suggested enterprise software systems development framework is the possibility of building a catalog of templates for heterogeneous enterprise software components based on an integrated metadata warehouse, i.e. a metadata snapshot repository. Therewith, software developers get a solution for storing relatively stable releases or frequently used configurations of heterogeneous enterprise software systems; such a metadata collection is often mission-critical in crisis.

Due to a database of the existing operational product releases, the approach potentially allows avoiding a number of integration problems related to combining "standard" enterprise software systems components. The approach allows substantial project development cost savings, provided the enterprise software systems developer's metadata snapshot repository already contains a solution similar to the new software product required. We recommend to enhance the metadata repository with the chronological sequences of enterprise product snapshots represented by trees, which contain the release baseline and its frequent variations stored as branches.

This is similar to version control CASE tools. For each software product and lifecycle phase, the approach allows a justified selection and cloning of most valuable deliverables in order to match different customers and/or different companies of the same customer.

In order to support uniform storage, search and cloning mechanisms, we recommend using problem-oriented domain-specific languages for business requirement specifications and the respective metadata [26].

Semantic-oriented storage, search and cloning mechanisms assist in quick reconstruction of the metadata snapshot from the repository components; the snapshot matches the updated product requirements in the best way. The approach potentially allows time-and-cost-effective transformation of the existing product components in order to match the new requirements, and thus is mission-critical in crisis. The approach is also suitable for strategic component reuse for the existing or new customers.

5   Large-Scale Systems: Pattern-Based Implementations

5.1   ITERA Oil-and-Gas Group: Enterprise Content Management

The process framework for the software product development included internet and intranet portal implementations at the ITERA International Group of Companies. The development included transformation of high-level formal specifications to UML diagrams by means of the ConceptModeller CASE tool. After that, the UML diagrams were transformed to target metadata snapshot schemes by means of CASE tools.

Based on the data model, the architecture was customized for enterprise content management. To provide the required industrial scalability and fault tolerance level, the integrated Oracle CASE toolkit was used to support software and business process re-engineering.

To provide uniform object-based content management in heterogeneous enterprise systems, a set of models was constructed. This set of models included the dynamic problem domain conceptual model and the state-based abstract machines for the enterprise environment and for the development tools.

To support the model set, a visual problem-oriented toolkit was developed for software system prototyping, design and implementation; the toolkit included ConceptModeller and the content management system (CMS) tools. According to the approach, an interface was designed for the internet portal. The interface was based on content-oriented architecture with explicit division into front-end and back-end sides. Portal design scheme was based on a set of data models, which integrated object-oriented management methods for data and metadata. We use the term content for the data and metadata together. Under metadata, we also mean knowledge. The major portals implemented in the ITERA Group were: CMS for network information resources, official Internet site, and enterprise Intranet portal.

To represent and manage high-level patterns, we used domain-driven development and DSLs [26–29]. The DSL-based approach helped to conquer problem domain complexity, to filter and structure the problem-specific information. It also provided a uniform approach to data representation and management. We used an external XML-based DSL, which extended the scope of the programming language for the enterprise software products.

5.2   Distributed Retail Network: Domain-Driven Messaging

A trading corporation used to operate a proprietary Microsoft .Net-based messaging system as an interface between its headquarters and local shops.

The system was client-server based. The client side included a local database and a Windows-based messaging service; the server side consisted of a Web service and a centralized database. One operation and maintenance challenge was complicated client-side code refactoring; it required frequent and massive recompiling and reinstallation. Due to highly coupled, non-flexible and non-transparent architecture, it was also difficult to detect, localize and fix product defects. Because of frequent code updates, the documentation was often inadequate to the current product release. Due to distributed and non-transparent system administration, remote shops configuration, product monitoring and management were decentralized.

The methodology implementation included detecting DSL scope, modeling problem domain, and development and testing of the DSL; the testing included DSL constraints verification.

Since the client side was the most changeable part of the product and the key challenge, the approach was client side-focused. The lifecycle model was iterative. The new product required a fundamental redesign of the architecture pattern (see Fig. 1). The fixed part of the application was a Windows-based host service; it contained a DSL parser. The input for the DSL parser was the transfer map for the current message.

For the flexible part of the problem domain, a DSL was developed. The DSL included rules and parameters for message transfer; it also allowed adding new message types. The semantic model included shops with different configuration instances, which represented the client-side message processing and transfer structure.

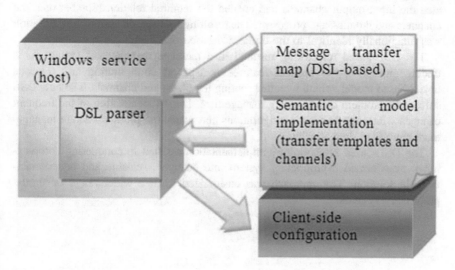

Fig. 1 MES architecture redesign

The next methodology stage was building semantic model for the DSL objects. We had three key types of the objects: messages, message transfer channels and message transfer templates. The DSL described object metadata, which included configurations and manipulation rules. Message templates were the core elements of the model, and channels were the links between the template instances. Templates and channels together made messaging maps. DSL described the maps, i.e. the static part of the model; messages were related to system dynamics and maintained the state.

The templates defined possible actions with the messages, including their transformations and routing. Templates were grouped together under a common interface. The standard routing templates were: content-based router, filter, receiver list, aggregator, splitter and sorter. We also produced a number of domain-specific templates for system reconfiguration, server interaction, and the similar operations.

For message management, we used channels. In the map messaging graph, templates were represented as nodes, and channels were the arcs between the templates. We implemented two types of channels: "peer-to-peer" channel and error message channel.

Based on DSL class model and implementation, we built the messaging maps; we used them to generate the product configurations by the DSL parser. At this stage, we built DSL syntax and semantics. A file instantiated each messaging map, which generally was a script. Each messaging map was an XML document; it defined product configuration and contained templates for routing and message processing. Each messaging map contained transfer channels and their relationships.

After parsing each messaging map, the parser created channel objects based on DSL channel descriptions. Then the parser configured the messaging system by creating message-processing objects in a similar way. Finally, the parser instantiated the input-output channels and created the required relationships between the channels and the message processor. The resulting DSL-based system configuration was functionally identical to the initial, C#-based one.

The DSL-based refactoring resulted in a message management system for a trading enterprise; the product featured transparent configuration and rigorous object-based model, which included routing templates and channels. The new DSL solved the problem of messaging management. Due to localization of the frequent changes within the transfer configurations and maps, the product change management simplified dramatically.

The DSL-based methodology implementation assisted in conquering complexity; it transformed a proprietary system into an open, scalable and better maintainable solution. The approach was customized to fit a wider range of similar proprietary systems.

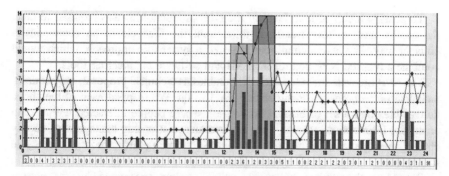

Fig. 2 The TAXXI application GUI

5.3   Air Traffic Control: Uniform Data Management

The problem was to develop remote access to the planning data for the air traffic control system. An operating solution already existed. However, it was based on an outdated TAXXI-Baikonur technology, which no longer evolved after early 2000s. The technology involved component-based visualized assembling of the server application. The standard Borland VCL library components were integrated with proprietary TAXXI components. The client side was TAXXI Communicator, i.e. a thick client.

The TAXXI technology was limited to Microsoft Windows platform, which was the only possible basis for client and server-side applications. According to the Russian State Program of Planning System Updates, the Main Air Traffic Management Centre planned to create a new remote access solution. The new internet-based architecture was planned as a Java technology-based solution to operate on the Apache web server platform. The solution was to query Oracle-based datacenter, process the query output and retrieve the results of the air traffic planned capacity to a user-friendly GUI client.

The practical application of the solution was the global enterprise-level software product, which was to provide uniform data access to all international air traffic participants (see Fig. 2).

Similar globalization processes were underway in Europe and the USA.

The idea of the pattern-based component approach was to unify the architecture-level updates and application migration, and to cope with the integration challenges of the global software product for air traffic management.

5.4   Nuclear Power Plant: 6D Modelling

Another implementation of the methodology was software re-engineering for nuclear power plants (NPP) based on high-level product patterns.

To provide a competitive level of the NPP production, it was required to meet the quality and security standards throughout the lifecycle in an economically efficient way. The above conditions required a systematic approach, which would combine advanced NPP control methods and software engineering tools.

The NPP lifecycle included such deliverables as technical proposal, project draft, technical project, design documentation and so on. We mapped each of the above deliverables into a set of business processes, where people interacted with certain enterprise information systems, such as customer relationship management, supply chain management, enterprise resource planning and product lifecycle management.

Based on the typical operation activity chains prescribed by the above-mentioned enterprise information systems, we formed automation standards for the NPP business processes. For example, workflow mechanisms assisted in building enterprise standards on electronic documents validation and approval. The enterprise systems acquired and maintained the data on the NPP lifecycle including such aspects as design, technology and economics. The combinations of business objects, such as 3D images, technological data and bank accounts, in the above systems described NPP as a large-scale object. Heterogeneous nature of these data objects and a huge number of production units made NPP an information object of high complexity.

A key competitiveness criterion in nuclear power industry was a set of the electronic manuals, which helped to assemble, troubleshoot and repair the NPPs. Such a set of manuals provided transparent information models of NPPs and their units, which allowed getting the required information on the object without directly querying it.

The 6D model included a combination of 3D unit schemes, time and other resources required to operate the NPP. Since the mechanisms for data object searching, filtering and associating were to provide complete and non-contradictory results, the information models required well-defined semantics. To provide unique data entries, the NPP software systems were to allow data acquisition throughout the lifecycle.

A single software product, such as a CAD system, could provide sufficient and adequate data for a relatively simple information, such as a 3D model. However, the 6D model required a combination of models from a number of systems, and its complexity could result in a local software development crisis. The methodology for building a 6D model suggested portal-based system integration, which was based on a software platform capable of the entire lifecycle support; this was Siemens Teamcenter.

Further development of the information model involved monitoring the entire system state and its influence on the other parts of the system. This helped to immediately respond to the critical issues in NPP construction. An example of such an issue was late delivery of a certain NPP unit, which prevented further construction activities. Such response agility was essential for decision-making, especially in crisis. Another source of local crisis, which might result in making a wrong decision was incomplete or incorrect information; the 6D model supported this case as well.

Constructing a nuclear reactor optimized for certain operating conditions, particularly from scratch is often a serious challenge. However, we managed to apply the above approach by constructing the product baseline from typical invariant units, and by adding a family of slightly varying branches on top of the baseline. Applying the methodology to the 6D information model of the nuclear reactor was a novel approach to pattern-based component development of NPP series.

6   Conclusion

In crisis, resource efficient software production is mission-critical. The chapter suggests a methodology of pattern-based software product development, which includes a set of formal models, processes, methods and CASE tools.

One key idea is resource efficient component development based on high-level architecture patterns for frequently used combinations of large-scale product modules. The other idea is using such pattern combinations as baselines and adding branches to them in order to manage requirement changes, release updates and similar requests from different clients. One more consideration is pattern development by means of domain-specific languages and visual CASE tools.

Lifecycle management for software development is a challenge in case of large-scale distributed heterogeneous applications. To solve the challenge, a uniform incremental software development methodology is suggested, which includes models, methods and supporting CASE-level tools.

The methodology implementations are particularly efficient in case of large-scale heterogeneous products, such as governmental organizations and commercial enterprises. In this case, the approach proved essential project time-and-cost reduction and an industrial level of quality.

We used the methodology to integrate heterogeneous enterprise components, including ERP modules for financial management, a legacy HR management system and a weak-structured multimedia archive. The portal architecture for the heterogeneous content provided a number of successful implementations in ITERA International Group of companies, which employed nearly 10,000 employees in over 20 countries.

The other areas of implementation included air transportation, retail network and nuclear power generation. Each implementation required development of a domain-specific language, which helped pattern-based product cloning and maintenance by means of specific CASE tools for data analysis and re-engineering.

The approach helped to identify the high-level pattern series of similar implementations; it resulted in term-and-cost reduction of 30 % and more. In crisis, we recommend to apply DSLs and CASE tools for these high-level patterns to the subsidiaries of existing clients and to new customers.

References

1. Wolfengagen, V.E.: Applicative Computing: Its Quarks, Atoms and Molecules, 62 pp. JurInfoR, Moscow (2010)
2. Naur, P., Randell, B. (eds.): Software engineering: report on a conference sponsored by the NATO science committee, 231 pp, Garmisch, Germany, 7–11th Oct 1968, Brussels, Scientific Affairs Division, NATO, Jan 1969
3. Schach, S.R.: Object-Oriented and Classical Software (8th edn.), 688 pp. McGraw-Hill (2011)
4. Sommerville, I.: Software Engineering (8th edn.), 864 pp. Addison-Wesley (2006)
5. Cook, S., Jones, G., Kent, S., Wills, A.C.: Domain-Specific Development with Visual Studio DSL Tools, 524 pp. Pearson Education, Inc. (2008)
6. Zykov, S.V.: Enterprise content management: theory and engineering for entire lifecycle support. In: Proceedings of CSIT'2006, Ufa State Aviation Technical University, USATU Publishers, Karlsruhe, Germany, vol. 1, pp. 86–92 (2006)
7. Zykov, S.V.: Integrated methodology for internet-based enterprise software systems development. In: Proceedings of WEBIST 2005, Miami, FL, USA, pp. 168–175, May 2005
8. Zykov, S.V.: An integral approach to enterprise content management. In: Callaos N., Lesso W., Zinn C.D., Zmazek B. (eds.) Proceedings of 11th International World Multi-Conference on Systemics, Cybernetics and Informatics (WMSCI 2007), Orlando, FL, USA, vol. I, pp. 212–216, 8–11 July 2007
9. Zykov, S.V.: ConceptModeller: Implementing a semantically-based toolkit for enterprise applications. In: Proceedings of CSE-2006, Lviv Polytechnic National University Publishers, Lviv, Ukraine, pp. 23–26, Oct 2006
10. Zykov, S.V.: Integrated methodology for internet-based enterprise software systems development In: Proceedings of WEBIST 2005, Miami, FL, USA, pp. 168–175, May 2005
11. Zykov, S.V.: An integral approach to enterprise content management. In: Callaos N., Lesso W., Zinn C.D., Zmazek B. (eds.) Proceedings of 11th International World Multi-Conference on Systemics, Cybernetics and Informatics (WMSCI 2007), Orlando, FL, USA, vol. I, pp. 212–216, 8–11 July 2007
12. Zykov, S.V.: The integrated methodology for enterprise content management. In: Proceedings of 13th International World Multi-Conference on Systemics, Cybernetics and Informatics (WMSCI 2009), pp. 259–264. Orlando, FL, USA (2009)
13. Zykov, S.V.: ConceptModeller: A frame-based toolkit for modeling complex software applications. In: Baralt J., Callaos N., Chu H.-W., Savoie M.J., Zinn C.D. (eds.) Proceedings of the International Multi Conferences on Complexity, Informatics and Cybernetics (IMCIC 2010), Orlando, FL, USA, vol. I, pp. 468–473, 6–9 April 2010
14. Kalinichenko, L.A., Stupnikov, S.A.: Heterogeneous information model unification as a pre-requisite to resource schema mapping. In: D'Atri, A., Sacca, D. (eds.) Software Systems: People, Organizations, Institutions, and Technologies. Proceedings of the 5th Conference of the Italian Chapter of Association for Software systems (itAIS), pp. 373–380. Springer, Heidelberg (2009)
15. Lenat, D., Guha, R.V.: Building Large Knowledge-Based Systems: Representation and Inference in the Cyc Project. Addison-Wesley (1990)
16. Masters, J., Güngördü, Z.: Structured knowledge source integration: a progress report. In: Integration of Knowledge Intensive Multiagent Systems. Cambridge, MA, USA (2003)
17. Nishizawa, H., Fujiwara, M., Yokoyama, M., Kanazawa, S.: R&D trends for future networks in the USA, the EU, and Japan. NTT Tech. Rev. 7(5), 1–6 (2009)
18. Lenat, D., Reed, S.: Mapping ontologies into cyc. In: Proceedings of AAAI 2002 Conference Workshop on Ontologies for the Semantic Web. Edmonton, Canada (2002)
19. Witbrock, M., Panton, K., Reed, S.L., et al.: Automated OWL annotation assisted by a large knowledge base. In: Workshop on Knowledge Markup and Semantic Annotation at the 3rd International Semantic Web Conference (ISWC 2004), pp. 71–80. Hiroshima, Japan (2004)

20. Zykov, S.V.: The integrated methodology for enterprise content management. In: Proceedings of 13th International World Multi-Conference on Systemics, Cybernetics and Informatics (WMSCI 2009), Orlando, FL, USA, pp. 259–264, 10–13 July 2009
21. Scott, D.S.: Lectures on a mathematical theory of computations, 148 pp. Oxford University Computing Laboratory Technical Monograph, PRG-19 (1981)
22. Barendregt, H.P.: The lambda calculus (rev. ed.) Studies in Logic, vol. 103. North Holland, Amsterdam (1984)
23. Curry, H.B., Feys, R.: Combinatory Logic, vol. 1. North Holland, Amsterdam (1958)
24. Wolfengagen, V.E.: Event driven objects. In: Proceedings of CSIT'99, Moscow, RussFiga, pp. 88–96 (1999)
25. Roussopulos, N.D.: A Semantic Network Model of Databases. Toronto University (1976)
26. Zykov, S.: Pattern development technology for heterogeneous enterprise software systems. J. Comm. Comput. 7(4), 56–61 (2010)
27. Fowler, M.: Analysis Patterns: Reusable Object Models, 223 pp. Addison-Wesley (1997)
28. Forbus, K., Birnbaum, L., Wagner, E., Baker, J., Witbrock, M.: Combining analogy, intelligent information retrieval, and knowledge integration for analysis: a preliminary report. In: 2005 International Conference on Intelligence Analysis, McLean, Virginia, USA (2005)
29. Evans, E.: Domain-Driven Design: Tackling Complexity in the Heart of Software, 560 pp. Addison-Wesley (2003)

Chapter 5
Knowledge Transfer: Manageable in Crisis?

Abstract The chapter discusses crisis-oriented patterns and practices for software development and knowledge transfer. We use a set of models to represent and manage the transfer; these are based on informing science framework and specific learning principles. We use process, data and system perspectives to describe software architecture. Our case studies include knowledge transfer from the Carnegie Mellon University masters' program in software engineering to the Russian Innopolis IT University. The CMU curriculum is integrated with project-based software development; the focus is quality and process improvement. The knowledge transfer requires special instructor training in teaching proficiency. We identified the key factors, which may inhibit technology transfer; these are differences in culture, geography and process maturity. To promote technology transfer in crisis, we suggest using multiple contexts and scaffolding, reducing overall cognitive teaching-and-learning load, establishing efficient feedback and mastering "soft" skills, such as communication, negotiations and time management. We recommend crisis-efficient metacognitive self-directed learning and addressing higher levels of Bloom's taxonomy including justification, analysis and practical application. As a result, the students learn to act as software engineering professionals and experts. Multi-context education includes hands-on learning-by-doing, "wicked" problems and "soft" skills required for crisis agility, which includes self-assessment, self-justification and self-adjustment. We also suggest a model for the knowledge transfer, which includes informing framework of transmitting and receiving sides, and the environment. For crisis, we recommend adding signal amplification, bidirectional feedback and amplitude thresholds. In crisis, successful knowledge transfer requires special training of the receiving side, which involves a number of human-related factors, such as prior knowledge, knowledge organization, feedback, mastery and practice. To ensure the learning quality, adequate bidirectional feedback-driven meta-cognitive cycle organization is very important. Both the students and the faculty should act as software engineering professionals and use their own expert-level judgment. Knowledge transfer should be multi-context and include hands-on project practice. Teamwork, communication and time management are the mission-critical "soft" skills in crisis environments, which are often complex and uncertain. Further, we suggest a high-level design

© Springer International Publishing Switzerland 2016

S.V. Zykov, *Crisis Management for Software Development and Knowledge Transfer*, Smart Innovation, Systems and Technologies 61, DOI 10.1007/978-3-319-42966-3_5 pattern approach in order to model the architectures for systems-of-systems; the model includes five application levels and two data levels. We instantiate this pattern by high-level architecture outlines for systems-of-systems in oil-and-gas and nuclear power industries.

Keywords Knowledge transfer · Informing science · "Soft" skills · System-of-systems

1   Introduction

In the previous chapters, we discussed lifecycle models and methodologies of the software product development. Software development is challenging due to a number of factors, including cultural issues, organization maturity level and mentality. However, the root cause of these key factors is human nature. Therewith, we argue that the human factor-related problems, specifically communication ones, are mission-critical, especially in crisis. The lack of common vision for the client and developer sides is often the source of crisis. To solve this problem, we should focus on the issues related to knowledge and technology transfer in software engineering-related environments, which is the topic of this chapter.

We are going to combine several models, and to apply them to a number of case studies (Fig. 1). One of the models we use for knowledge transfer is known as Cohen's informing framework approach [1]. The informing framework usually names the client and the developer as sending and receiving sides, and the environment as a medium.

The other knowledge transfer model is based on a set of learning principles and involves metacognitive teaching-and-learning [2].

The case studies for knowledge transfer include a number of software engineering-related environments, such as large-scale, mission-critical software product development and master curriculum development for a leading IT university.

To analyze the human-related factors in the context of large-scale software development, we enhance the multi-level process framework introduced earlier with the views for data and software products, which make the IT infrastructure. We arrive to a matrix, the columns of which correspond to processes, data and systems, and the rows of which contain levels, such as hardware, database, data acquisition, daily management, strategic planning and decision-making. We call this the enterprise engineering matrix and discuss how we can use it to detect and predict local crises in the areas of oil-and-gas industry and nuclear power production.

This chapter is organized as follows. Section 1 presents an outline of the environment and framework for the knowledge transfer. Section 2 introduces the informing framework application to knowledge transfer modeling. Section 3 identifies the human-related factors, which influence the knowledge transfer. Section 4 describes the new project of a Russian IT city, which we use as a case

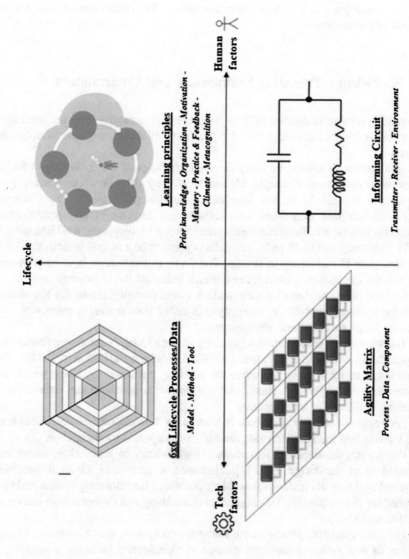

Fig. 1 The process, data and system models organization

study; Section 5 addresses the identified human-related factors for this new project. Section 6 revisits the high-level design patterns in the case of enterprise software development in crisis. Crisis-agile examples include an enterprise engineering matrix, a process knowledgebase and a layer-based approach to process management. We further instantiate this layer-based approach by software system outlines for an oil-and-gas and a nuclear power enterprises. The conclusion summarizes the results of the chapter.

2 Knowledge Transfer: Framework and Environment

In crisis, a number of factors, such as cultural issues, maturity level and mentality may help or hinder knowledge transfer. All these key factors originate from human nature.

This section describes the early steps and findings of the knowledge transfer from world-renowned Carnegie Mellon University (http://www.cmu.edu/) to a brand new Russian IT startup, Innopolis (http://innopolis.ru/en/). The Innopolis City will integrate experience of academicians, researchers and practitioners. Innopolis will be a self-sufficient ecosystem with a kindergarten, a STEM school, an IT University and an IT park. Innopolis is going to be a unique project, since it is the first pure IT university in Russia. The focus of the section is to identify and discuss the key human-related factors, which influence the knowledge transfer.

Clearly, plain copying of the curriculum would probably hinder the knowledge transfer because Innopolis is an ecosystem rather than merely a university, and because of its cross-cultural environment.

The knowledge transfer model generally follows Cohen's informing framework approach [1], which, in turn, is based on Shannon's communication model [3]. The knowledge transfer model identifies the sending and receiving sides, which are curriculum developer and client, and the medium, which is the cross-cultural teaching-and-learning environment.

An important observation, which holds true for the Innopolis project, is that both the sending and the receiving side are informing systems by themselves [4].

Curriculum development represents a higher level of abstraction, sometimes referred to as the design level [4]. Therewith, a meta-level model is generally required to divide the multi-level problem domain of the informing system and thus to conquer its complexity. This higher-level modeling and design is also known as architecture [5–7].

For such complex systems as the Innopolis ecosystem, interrelatedness, i.e. the ability of one system component change to significantly influence a number of adjacent components changes, is critical [7].

We are going to analyze the knowledge transfer lessons learned from our training in software engineering at Carnegie Mellon University. The training purpose was master's curriculum development for the Innopolis University.

First, let us introduce the mission-critical aspects of the knowledge transfer environment in a more detailed way; this will help us to understand the complexity and the crisis nature of the problem.

For a number of years, the Russian government plans to increase productivity growth [8]; however, the Russian IT sector still faces a number of challenges. First, the existing technological infrastructure is often inadequate. Second, the intellectual property laws are generally weak, and the regulatory environment is problematic. Third, public and private sectors do not collaborate resonantly. Fourth, skilled professionals number is insufficient and, consequently, hiring skilled IT employees is a challenge. Fifth, major IT companies report the quality of the university education as insufficient.

In 2010, Russia started the Innopolis project to solve the above problems. The new IT city location is the middle part of Russia, near Kazan', the capital of Tatarstan Republic. Kazan' is some 1,000 miles to the Southeast from Moscow, the Russian capital. Federal Russian and local Tatarstan governments support the ambitious project. The name Innopolis means "Innovation City" in Greek. The IT ecosystem will include the Innopolis University (Fig. 2), which is to provide the Russian IT industry with education, and research excellence center [9]. The expected results are nearly two-fold increase in innovative industry implementations and substantial growth of GDP and hi-tech goods production.

The key idea of the Innopolis project is to bring together education, research and practice, and to create a unique IT city for state-of-the-art software development and recreation facilities. Currently, the project is in its pilot stage (Fig. 3).

Fig. 2 Innopolis University (*photo* by the author)

Fig. 3 Innopolis University interior (*photo* by the author)

By 2030, the new city will host around 150,000 inhabitants. The Innovative City will become Kazan's satellite—the capital of Tatarstan is some 20 miles away. A big challenge is that the entire Innopolis infrastructure started literally from scratch: the nearby surroundings are currently uninhabited. The other critical parameter is the management, which should compromise the objectives and goals of the multiple participating sides: governmental, industry, academic and research organizations. However, the initial idea to integrate the powers of the two governments, IT academicians, researchers and practitioners in a single location will likely assist in coping with these challenges.

Fig. 4 Innopolis IT park building (*photo* by the author)

Innopolis will support the entire lifecycle of IT people; the city already has a kindergarten, a STEM training school, a university and an IT park (Figs. 3, 4 and 5). Of these, our primary focus is the university, since it is the brainpower of the new city.

Fig. 5 Innopolis residential facilities (*photo* by the author)

Fig. 6 CMU: Institute for Software Research (*photo* by the author)

Specifically, let us discuss the master curricula in software engineering and the takeaways from CMU master curricula in software engineering and IT (Fig. 6). The idea is to identify the ways to adjust the CMU curricula to the Innopolis environment.

Fig. 7 CMU: Institute for Software Research, interior (*photo* by the author)

CMU has been chosen for Innopolis faculty training since this is the birthplace of the software engineering and a top rank university in SE and IT [10]. The CMU software engineering master's program is a well-balanced alloy of research and project practice, core and elective courses. However, direct curricula copying from CMU to Innopolis is probably not an entirely correct solution because these two universities obviously have certain environmental differences (Figs. 6 and 7).

3   Modeling the Transfer: Informing Framework Approach

To model the knowledge transfer, we will use informing framework model and the concept of resonance [4]. In terms of the informing framework model, the knowledge transfer process should be optimized in order to make it more resonant. As Gill and Bhattacherjee state, "... without knowing the client, predicting resonant communication forms is impossible" [4]. Thus, building a large-scale and complex Innopolis ecosystem literarily from scratch is certainly a dramatic challenge.

As we know from the electronics, a basic LC circuit, which consists of capacitor and inductive coupling, uses feedback to make it oscillate (Fig. 8).

Within the LC circuits, however, there are two kinds of feedback, which are positive and negative, and the distinction between these two is critical. The so-called positive feedback increases the gain of an amplifier, while the negative feedback reduces the gain. However, both types of feedback require wise application.

Careless application of the positive feedback is in fact the source of the most common problem with the amplified circuits, since it causes unwanted spontaneous oscillations. Possible consequences are an amplifier, which may produce an output

Fig. 8 Basic LC oscillator circuit

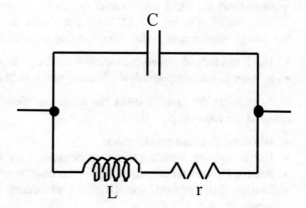

Fig. 9 LC oscillator circuit
with positive and negative
feedback paths

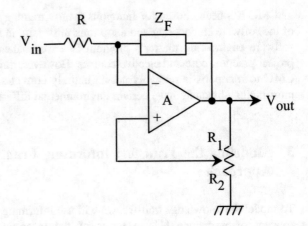

with no input, and even destruction of the circuit due to spontaneous resonance. Similarly, careless application of the negative feedback can decrease the gain so that regardless of the input value it either fades down to a negligible output value or produces no output at all.

The above considerations suggest that the informing circuit should be balanced, so that a carefully selected combination of positive and negative feedback loops produces controlled resonance at any given point of time (Fig. 9).

The understanding of the causes of spontaneous oscillation, which is usually unwanted, is essential for designing and "debugging" the feedback loops within the informing circuits. On the other hand, positive feedback has its uses. Essentially, all sources of the informing signals contain oscillators that use positive feedback. Positive feedback is useful in the circuits that determine whether an informing signal value exceeds a threshold, even in the presence of a noise. The noise may come from the transmitting side, which has a complex structure. The noise may also originate from the client side network, which is also complex.

The idea of applying the LC oscillator circuit theory to the knowledge and technology transfer area is that the informing "signal" should:

• Use a carefully designed bidirectional feedback circuit;
• Remain within the prescribed operating range of lower and upper thresholds.

To promote the transfer under the crisis conditions identified above, we recommend the following:

• Introduce clear cut ground rules;
• Use transparent, unambiguous transfer policies and strictly follow them;
• Reduce the transfer cognitive load to the least amount possible;
• Provide fast, frequent, specific and goal-directed feedback for the transfer lifecycle;
• Establish and maintain open, warm, friendly environment or "climate" [1, 4].

4   Knowledge Transfer: Detecting Human-Related Factors

After CMU visiting faculty training, we detected an interrelated set of key factors that influence the knowledge transfer. We interpret these factors in terms of "seven principles" [2], which "inform" the knowledge transfer processes. In our view, the mission-critical factors to consider in crisis include prior knowledge, knowledge organization, mastery, feedback and practice, course climate, and motivation. Let us analyze the nature and impact of these factors on the knowledge transfer between CMU and Innopolis.

The role of prior knowledge in the software engineering education at CMU is not that straightforward as it is in a number of other domains. Oftentimes, particularly in the Software Architectures course, the CMU students relate what they learn to what they have known previously. Therewith, they often interpret the new knowledge using prior knowledge, beliefs and assumptions [11]. Though many researchers agree that students usually connect their prior knowledge to the new knowledge in order to learn, the knowledge generated by the students within the IT domain may often be incomplete or inappropriate [12]. In our view, a positive way to approach this challenge in new and prior knowledge is to self-reflect and to use case-based reasoning. Research also suggests that mentoring the students through a process of analogical reasoning helps them to focus on deeper problem domain relationships rather than on superficial similarities, which often are quite tricky [13]. After analysis of multiple business cases, the students tend to build more effective knowledgebase and learn more efficiently [14].

Another human-related factor, which can substantially influence the knowledge transfer processes, is self-motivation [15]. However, we should make a clear distinction between the subjective value and the expectations for successful attainment of the goal [16]. We should also notice that the students with learning goals, as opposed to performance goals, often use complex strategies for deeper mastering the curricula and conquering challenges [17]. To develop mastery in software engineering, the students should:

- Acquire component-based skills, such as architectural diagramming and quality measurements;
- Practically integrate these component-based skills, for example, by analyzing architecture trade-offs in software quality attributes;
- Consider the applicability constraints, such as customer and stakeholder priorities of business and technical nature.

The mastery human-related factor changes from novice to expert level; it includes issues of competence and consciousness [18]. While acquiring mastery in software engineering, the students seldom succeed in direct application of their own skills to a different context; however, after thoughtful mentor prompts, which may seem to the students subtle at first sight, they usually succeed [19, 20].

Practice and feedback are the closely interwoven factors; they are critical for professional-level balance of theoretical knowledge and practical skills. To improve

Fig. 10 Training lifecycle
diagram with ordinary
cognitive load

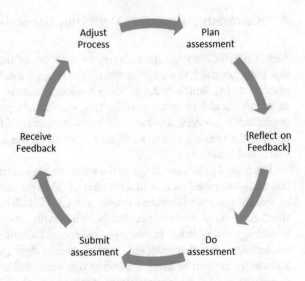

knowledge transfer, we recommend focusing both factors on the same learning objectives [2]. The essential qualities of feedback are frequency and timeliness; these suggest a clear trade-off with feedback size. However, if we compromise on the feedback size, a better knowledge transfer is often possible [21]. Feedback, together with individual mentoring, often helps to set the appropriate level of assignment difficulty, and thus to promote knowledge transfer [11].

To fully utilize self-directed learning, which is another mission-critical factor, the students should efficiently implement their own metacognitive cycle (Fig. 10). They should evaluate their own knowledge and skills, plan and monitor progress, and adjust strategies [22]. For successful knowledge transfer in crisis, the students need to actively use their metacognitive skills [23]. The key factor to improve the metacognitive skills is self-reflection; it helps to continuously monitor and adjust the problem-solving progress. To improve software engineering-specific metacognitive skills, we recommend to use a set of related techniques, which includes team-based concept mapping and brainstorming [24].

5 The First Ever Russian IT City: Facilitating the Transfer

The Innopolis project started in 2010. Russia has a long academic tradition; however, the project is a challenge in itself. This is so because prior to Innopolis, none of the Russian universities was an entirely IT school. In contrast to classical Russian universities focused on basic research, the Innopolis is to specifically aim at industry-focused IT applications. Innopolis city is compact; the facilities are within a 15 min walking distance, and the living standards are world competitive.

Innopolis location is really unique: Kazan' city, the Tatar capital, is located just between Europe and Asia with an average flight time to major European and Asian destinations of nearly 1.5 hours shorter than that from Moscow.

The location is international: Kazan' city has over 1,200-year history [25], it is world famous for academics and research. The dominating sectors are oil-and-gas, automobile industry, construction, transportation and communication. Recently, the city was renovated to host a number of major international events. Large IT multinationals, such as Fujitsu-Siemens, are located there. The first stage of the project is complete. It includes the main university building, residential halls and the IT park offices; the current capacity is 5,000 people.

The university is the Innopolis brainpower; thus, knowledge transfer issues are mission-critical.

Let us consider the source of the knowledge transfer, the CMU software engineering master program in more detail. The CMU master in IT and software engineering (MSIT/SE) training schedule is very busy, which is a potential source of knowledge transfer crisis (see Table 1). The curriculum lasts for exactly one calendar year; however, it is divided into three semesters: fall, spring and summer. Each of the three semesters is nearly 4 months long; the midterm breaks are about 1 week short [43, 44].

The CMU MSIT/SE core courses are tightly interwoven; they are integrated with practical software development projects.

The primary focus of the visiting faculty training was the Architectures for Software Systems. The course is case study-based; it has two versions: one is for master students, the other is for enterprise executive training.

The course focus is the change of chaotic software product development for process-driven one. The general idea is to have a "just right" amount of the formal processes to be efficient; this is similar to the crisis optimization concepts, which we discussed in Chaps. 2 and 3. According to the course, the chief architect adjusts the process to fit the project aims and scope. The course is practically oriented; it is

Table 1 CMU MSIT/SE required courses	Course code	Course name
	17-602	Introduction to personal software process
	17-651	Models of software systems
	17-652	Methods: Deciding what to design
	17-653	Managing software development
	17-654	Analysis of software artifacts
	17-655	Architectures for software systems
	17-656	Communication for software engineers I
	17-657	Communication for software engineers II
	17-677	MSIT project I
	17-678	MSIT project II

supported by recitations and driven by a real-world project, which has an actual customer and an actual software product as an outcome. Teamwork on these real-world projects is also known as Studio projects; it is aligned with the individual assignments.

Perhaps, sometimes the individual contribution of each student to the team project is not quite clear, at least in terms of the final grade, which is heavily dependent upon team results, so certain "free riding" is quite possible. However, getting a high grade is doubtful for any "free rider", since individual contribution amounts to around 30 % of the final grade. Overall, the course grading policy is fair.

The course can scale up to large projects; however, Studio project teams are small, are their software products are typically of small and medium scale. The ACDM/ATAM framework fits the Studio projects; however, it is often a challenge for the students to adjust the basic framework in order to fit the project size and scope [5]. For a master's level course in software engineering, this challenge is a strong point; it teaches the students to think architecturally and systemically.

Another strength of the course is the "just enough mathematics" approach. On the one hand, the course assignments never violate mathematics. On the other hand, this approach keeps the course focused on its primary goal, which is to form the architectural thinking.

It is a challenge to master the course. One reason for this is that it requires intuition and influences the way of thinking; these skills are hard not only for students, but also for some software engineers.

The other MSIT/SE core course examined by the visiting faculty was the Analysis of Software Artifacts. The course is comprehensive; it embraces a number of areas including quality planning, model checking, testing techniques and a few others. Similar to the Architectures, the Analysis course approaches software quality systematically and focuses on practices and techniques for quality assurance. Again, there is a number of courses for software testing, which focus on specific techniques, such as "white box", "black box" and state-based testing. The CMU Analysis course also deals with the above techniques to a certain extent; however, its overall goal is much broader and more complex. The course introduces the idea of software quality improvement throughout the entire lifecycle. It is not just coding and testing which add to product quality; the entire development process matters. That is why the course includes a CMMI maturity model overview among the other standards [7].

Another issue, which contributes to software quality, is security. Again, this is a wide area; however, security relationship to software quality often remains somewhat subtle to the students.

Systematic approach to quality through product security testing adds value to the course. The CMU Analysis course systematically presents combinatorial testing, queuing theory, defect taxonomies and a few other aspects; this makes the course outstanding. The course gives a brief overview of quality assurance, and the students are supposed to know the key dates and names. The course is supported by a "launch pad", where the students can learn software quality assurance by actually doing it. The "launch pad" contains a large code repository with a certain amount of defects injected, and the students apply the quality assurance methods they have just studied in a realistic environment of "third-party" software artifacts.

Reference and reading materials for the CMU Analysis course are top quality. The course uses a database of publications, the size of which totals to gigabytes. The database is full-text searchable; it takes the teacher a couple of minutes to extract the required reference and several more minutes to incorporate a quotation, a table or a graph into a lecture slide. This is an impressive illustration of the quality of this innovative courseware. The course provides many helpful references for further reading and self-improvement concerning the above-mentioned areas; it potentially assists in postgraduate studies, research and publications.

The Analysis course fits into the CMU MSIT/SE framework of the Studio project; at a certain point of time, the project requires a quality plan of the software production. Each student team gets at least two attempts to present the intermediate results; certain techniques are used to engage every team not only into the presentation, but also into the discussion on the product quality. Thus, students get the understanding of how and why does the software product quality matter. Teamwork is aligned with the individual assignments. The Analysis course is a well-justified combination of individual and team activities; it is professionally monitored, mentored and assisted.

Grading policy of the Analysis course is transparent, explicit and fair. The rubrics are clear, concise and easy to use; feedback is instant. Due to properly organized processes and clearly stated "rules of the game", grading is fast and smooth; the students get a clear, well-justified and prompt feedback. Teaching assistants are involved in not only lectures, recitations and grading, but also in developing assignments and rubrics.

The "just enough mathematics" approach seems be the right choice for the Analysis and the Architectures courses. The Analysis course potentially requires sophisticated mathematics, such as specific forms of Petri nets, advanced statistics and the like. However, software engineering practitioners would seldom use these complex methods; moreover, the program has a dedicated course of Models of Software Systems. Still, since the engineering approach has to be rigorous and metrics-based, the Analysis course introduces and teaches how to apply certain metrics, such as throughput, and general laws, such as Little's law. The "just enough mathematics" approach keeps the students practically focused on product quality improvement.

One more MSIT/SE course examined by the visiting faculty was the Personal Software Process (PSP). The course precedes the core MSIT/SE program; it introduces a way toward a more efficient software development. The subject is not merely coding in terms of design and testing. Rather, it is process-based software development, which includes planning and monitoring the project progress. Planning and monitoring processes take time and effort; however, they pay back soon, and the course helps to apply this practically. The PSP is a predecessor of the CMM and CMMI standards [7, 54]. The course uses statistics in order to assess the current software development process, detect the bottlenecks, identify the improvement areas, adjust the project planning and monitoring, and follow the updated process. The course focus is continuous individual improvement, which is a value in itself. The basic course reading by W.Humpfrey illustrates how a disciplined software development pays back in short-term and long-term perspectives [26]. Forecasting is also a major course takeaway. The course, however, does not merely retell the book chapters. Instead, it features a toolkit for personal improvement, which includes code reviews, the Pomodoro technique for time management, and a number of other powerful practices. These tools build up the students' professional skills; they are extremely useful and helpful in their intensive master program studies.

The PSP course adds to the powerful framework of the Analysis and the Architectures, since it focuses on improvement of software product quality and project planning. At the team level, further process improvement is possible; however, since PSP focus is the individual process, the course addresses this issue in brief only.

The Software Process Dashboard tool supports the PSP course; the tool assists in planning, monitoring and estimating of the software product development [27]; it helps to analyze the process data.

The PSP course features tools and techniques for fast data extraction and analysis. Course grading templates are available to speed up the assessment process in case of large-scale student groups; remote delivery is a possible option. Early in the course, the students focus on the continuous improvement of their individual development skills; this often becomes a personal motivator and driver, and helps to manage crises at the individual level. The more accurately the students collect and analyze the data, the better they can estimate and forecast their future development results. The course provides a number of links to the state-of-the-art standards and policies of software development; this helps to align it with the future requirements, which the students will likely encounter.

The PSP course assists in the Studio projects and their outcomes; it provides document templates, planning strategies, and estimating techniques, which help students to justify their results and to adjust in order to cope with local software development crises.

The PSP rubrics are easy to follow, they work both face-to-face and remotely; course grading is efficient and accurate. Due to properly organized processes and templates and clearly stated "rules of the game", the students get a clear, justified and prompt feedback.

After discussing the core MIST/SE courses, let us summarize the key learning principles for the new Innopolis university software engineering program.

First, the learning process should include instant hands-on application of the principles learned into practice. This is likely to result in a mastery level comparable to that of a professional software engineer. The university curriculum should consistently link theory and practice by means of realistic business cases.

Second, the theoretical foundation should be just enough to support multi-context practical application. Third, instant and frequent feedback is required to practice the project-based skills, which are required to achieve mastery. Within the project framework, the teachers become mentors; they provide instant feedback. Team and individual mentoring is required in order to fine-tune the assignment difficulty level so that the students would be able to master the multi-discipline problem domain, which requires a number of "soft" skills. These "soft" skills are: management, communication and team-building, to name a few. Other essential skills include interpersonal and cross-cultural communication, negotiation, conflict and risk management, relationship building and maintaining, and similar abilities.

The project-based learning should provide students with teamwork, which adds to their interpersonal "soft" and core practical skills. The learning model should be based on continuous self-reflection, self-motivation, self-monitoring and self-directed adjustment. Self-motivation should include setting up personal goals and identifying the skills required to achieve them. Self-adjustment refers to students' personal flexibility to meet the product requirements of the stakeholders and teaching expectations of the professors in the master courses. Such self-adjustment adds personal agility, which is required to manage individual crises.

6   Knowledge Transfer: Addressing Human-Related Factors

The key human-related factors that we identified earlier are critically important for knowledge transfer, especially in crisis. Let us relate these factors to the observations and findings of the CMU visiting faculty training.

One of the human-related factors is prior knowledge; it may either help or hinder the knowledge transfer. In case of faculty training, the amount of prior knowledge on Architectures for Software Systems course was large enough. However, certain prior knowledge concepts were inappropriate and/or irrelevant; this significantly hindered the transfer. One example of the hindering prior knowledge was excessive focusing on real-life implications of the economic and business constraints, which were predominant in Russia, instead of merely justifying technologies as prescribed by the course objectives. Therewith, the knowledge transfer was incomplete, and the result was Guest Lecturer certification rather than Primary Instructor one. At CMU, the Guest Lecturer is usually responsible for 20 % of the course content or less. However, since Architectures for Software Systems course was generally beyond the primary domain of the visiting faculty expertise, the assignment of the Guest Lecturer status was perhaps the optimal solution.

Other human-related factors are knowledge organization and mastery. Concerning knowledge organization, experts often omit evident steps of reasoning, or make "shortcuts", to use the terms of knowledge receiving side. Due to a certain level of the visiting faculty mastery, the knowledge receiving side accepted this "shortcuts" policy by default.

The other course was Personal Software Process; it heavily focused on process definition and software development discipline. Due to the nature of the course, the knowledge transmitting side did not intend to accept any "shortcuts". Therewith, due to lack of the common vision of the sides, the first attempt of PSP training did not result in a satisfactory level of knowledge transfer and in a valid instructor certification. However, after a self-reflection by means of writing a final report, and after proper adjustment to the training side requirements, the knowledge transfer was successful, and the visiting faculty was certified as the Primary Instructor. Another critical success factor was rigorous following the process. Again, this was a requirement of the knowledge transmitting side. However, due to cultural diversity, the initial understanding of the knowledge receiving side set up a different level of process agility and transfer expectations, which resulted in certain certification issues for the initial course take. During the retake, however, there were certain process adjustments. These adjustments were based on a self-reflection of the knowledge receiving side and on a more frequent and goal-directed feedback from the knowledge transmitting side. These process adjustments resulted in a resonant knowledge transfer and eventually successful Primary Instructor certification.

Let us discuss the difference between the knowledge organization and mastery of the Guest Lecturer and the Primary Instructor. The Guest Lecturer is aware of the critical relationships between the key course concepts; however, his/her network of the course-related concepts is not as rich as that of the Primary Instructor. This may inhibit the initial path of the teacher-to-learner knowledge transfer so that the knowledge transfer may become somewhat inaccurate and/or inappropriate.

In case of Architectures for Software Systems, identification of the mission-critical software quality attributes would probably be a key to successful knowledge transfer. One realistic example of a mission-critical quality attribute was the availability of a robot-controlled warehouse. The knowledge receiving side was generally adequate on grading course assessments. However, a deep, mentor-level evaluation of the grading process by the knowledge transmitting side showed that delicate probing and hinting students on their own context-specific application of mission-critical concepts was problematic for the knowledge receiving side. A realistic example of such a context-specific application was a well-justified architectural framework for a robot-assisted warehouse.

The knowledge transfer can be divided into near and far [2]. The near transfer means knowledge applicability in similar or adjacent problem domains only, while the far transfer usually allows a multi-domain application in different contexts, some of which are substantially far from the original one. The question is how far the CMU-Innopolis knowledge transfer was possible. Our experience showed that there was hardly any far transfer evidence; however, a near transfer based on cloning of the curriculum with certain minor adjustments was possible. In our opinion, for each course the far transfer would require at least two attempts with a self-reflection, subsequent adjustment and multi-context application.

In case of Architectures for Software Systems, the rich prior knowledge was excessive in certain aspects, such as fault tolerance and data consistency, which inhibited the far transfer that was required for Primary Instructor certification. However, the same prior knowledge and complex knowledge organization in terms of extensive concept relationships promoted knowledge transfer in certain knowledge areas, such as enterprise and mission-critical systems, and resulted in Guest Lecturer certification.

Another lesson learned was that the educational framework and terminology, including the core readings for a course, influenced the knowledge transfer largely. As an example, the other trainee, a USA University Professor specializing in Software Architectures, was physically present at 30 % less classes. However, his prior knowledge and knowledge organization structure were much more relevant to these of the knowledge transmitting side. He had been delivering courses on Software Architectures in the USA for a number of years. His courses were based on the same core reading; this was a famous book written by three prominent Software Engineering Institute experts [28]. The Software Engineering Institute is located next to the Carnegie Mellon University, and the research activities of the two organizations are tightly interwoven. Therefore, the key course concepts from CMU and the other trainee were very similar, and the prior knowledge bases of both sides used similar sources. Moreover, concerning practical application of the concepts and principles of software-intensive systems design, the CMU teachers and the other trainee built their knowledge bases and business cases on similar real-world projects.

One more mission-critical human resource-centered factor for knowledge transfer is relevant practice based on timely feedback. In theory, practice and feedback require alignment [29]. Concerning the knowledge transmitting side, the feedback was not always ideally aligned with the assignment submission schedule. The primary reason for that was intensive mobility (due to a number of business trips) and non-synchronized vacation times of both sides. Another challenge was the feedback itself, which sometimes seemed too general or too vague for the knowledge receiving side. For the PSP course, the feedback was based on an automatically generated template for student submissions; a batch-processing tool assisted in it. Thus, a certain portion of the feedback was insufficiently targeted to address the issues of the knowledge receiving side and to link them to the prior assessments. Therefore, the knowledge transfer was inhibited.

Because of delayed start, the knowledge receiving side postponed the most intensive part of the training until the summer semester and switched to distant learning. Due to summertime, both teaching and learning sides were traveling intensively, so, the season and distance factors hindered timely and detailed feedback. The initial certification attempts for Architectures and PSP were not quite successful; therefore, the sides agreed on the strategy of sequential assessment resubmission.

Eventually, both retakes resulted in successful certification, which confirmed that the knowledge transfer did eventually occur. A more detailed feedback accompanied practical assignments for both course retakes.

Even though both course retakes were remote, the cognitive load essentially decreased. The steps to reduce the cognitive load were the following:

 (i)   less submissions "on-the-go", for example, while traveling;
 (ii) step-by-step training plan approved by both sides; this included realistic
       milestones and clear deliverables;
(iii) a well-justified amount of time invested into the course retakes;
(iv)   a detailed and timely feedback after every assessment;
 (v)   future submissions adjusted on the basis of the feedback for the previous one(s).

The training lifecycle diagram with reduced cognitive load is presented in
Fig. 11.

To reduce the cognitive load, the knowledge receiving side seized a number of
concurrent learning activities and focused on doing "one thing at a time" as much as
it was possible. In contrast to the previous attempt, the final report for the
Architectures course was separated from any other knowledge transfer-related
activities. A fair amount of time, nearly 100 hours in just 2 weeks, was invested
into the final report. Only after the final report submission for the Architectures
course, which strictly followed the deadline set by the knowledge transmitting side,
the PSP course retake started.

Thorough planning included certain time slots to revise materials, install, con-
figure and test the new development environment, and to do some other preparatory
activities. The Final Report on PSP, though it took approximately an order of
magnitude less time, was produced according to similar scenario than that on the
Architectures. Again, this reduced the cognitive load to minimum, and allowed
concentrating on course-specific issues, such as process planning and monitoring,
self-evaluation and adjustment (see Fig. 11).

The feedback on the Architectures course included a number of activities. These
were discussions on reading questions, grading individual and group assignments,
and writing the final report. As our estimates show, incomplete knowledge transfer

Fig. 11 Training lifecycle
with reduced cognitive load

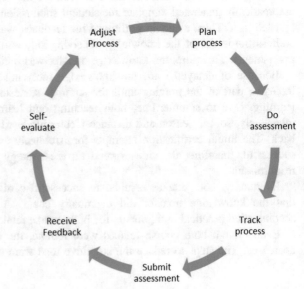

and, consequently, partial certification could result from insufficient feedback. As for the intermediate assignments, which were reading questions and individual/group practical activities, the feedback received was probably not always as prompt and goal-directed as it was desired by the knowledge receiving side. Another reason was a heavy cognitive load for the dedicated instructor of the knowledge transmitting side, which again was a crisis-related human factor. The instructor had a bulk of administrative load, due to which he had to arrange intensive online meetings and take frequent business trips. Additionally, an 8 hours time shift between Moscow/Kazan in Russia and Pittsburgh in the USA, and different national holiday schedules inhibited feedback efficiency. This was the key reason for a number of delayed meetings and late feedback emails, and consequently, for hindering the Architectures knowledge transfer.

There were three final report revisions for the Architectures. An exhaustive feedback with clear guidelines followed the first one. Based on the initial feedback, two more revisions of the final report followed; however, the feedback was not as detailed as before. Thus, feedback was probably the primary human factor that hindered the knowledge transfer process.

The outcome for the other certification course, PSP, was different. However, the primary human factor was probably also the feedback. For the initial attempt, a typical feedback for earlier submissions was several days late. This resulted in the knowledge receiving side misunderstanding of the ground rule that any further assignment was dependent on the previous one(s). However, the feedback that followed clarified that most of the previously submitted assignments did not meet that ground rule, and thus they were subject to resubmission.

To scaffold the further learning process and to promote knowledge transfer, the knowledge transmitting side suggested a context change, which included a different programming paradigm and a different programming language. At first, this could seem a misleading over-complication. However, the multi-context training resulted in a better knowledge organization scheme in terms of matching course-specific objectives. Moreover, this multi-context approach allowed scaffolding and organizing the conceptual scheme of the prior knowledge so that the representations of the knowledge transmitting and receiving sides became better aligned.

The PSP context change required certain preparatory efforts, such as installing a new integrated development environment and revising specific programming language concepts. However, the result was successful due to the following human-related factors: better meta-knowledge transfer, improved feedback quality and adjusted meta-cognitive cycle [38–41, 48]. The process improvements included certain updates in the training plan, certification requirements, submission ground rules, and a few other aspects. The feedback became faster, it followed the pattern of "one assignment at a time". The feedback focused on submission benefits; it allowed time gaps to analyze, self-reflect and adjust. Therefore, knowledge transfer appeared to be critically dependent on the feedback. Our general requirements for the instructor feedback are frequency, timeliness, focus, and goal-orientation. In our opinion, late feedback inhibits the knowledge transfer, and it may result in non-certification.

The multi-context approach was also applied to the Architectures course. The initial context was traditional MSIT/SE curriculum; the second one was LG company professional training. The second context had a realistic hardware-and-software system as a goal to implement, so it articulated the subtle theoretical concepts in a more accurate and multi-dimensional way. Thus, multi-context application of the key course concepts adjusted the knowledge organization scheme of the knowledge receiving side so that course specific issues and challenges were identified, highlighted and received scaffolding. Concerning the PSP course, similar multi-context effect resulted from applying structural and object-oriented development paradigms, Free Pascal and NetBeans environments, and Pascal and Java programming languages.

Another key human-related factor of the knowledge transfer is achieving certain level of mastery, i.e. a set of domain-specific knowledge, skills and attitudes. Concerning software engineering, building and transferring mastery involves a number of non-technical "soft" skills, such as business communication, teamwork, and time management [21]. Though these skills are non-core in terms of the discipline, they are critical for a successful knowledge transfer, and neglecting them may result in a crisis. However, the Russian software engineering academia generally underestimates the value of the "soft" skills, whereas the CMU faculty members explicitly focus on them. For cross-cultural contexts, these skills are mission-critical.

The aim of software engineering is product-based development and maintenance of large-scale and complex software systems; it requires efficient communication between the client and the developer. A software engineering product is a result of teamwork. Design, development, testing and maintenance teams often differ; however, they should collaborate efficiently to build and support a competitive product. Thus, teamwork skills are critically important for the knowledge transfer in any software engineering curriculum. In Chap. 3, we discussed open communication and self-managed teamwork issues in the context of software development methodologies.

Another human-related factor that hinders knowledge transfer is maturity level difference; this was more formal for the knowledge transmitting side and more ad hoc for the knowledge receiving side [30].

The other human-related factors, which may affect the knowledge transfer, are motivation and course climate. Though these are not critically important for faculty training, they still are influential in crisis. For example, late training start of the knowledge receiving side resulted in skipping at least 25 % of the classes at the beginning of semester. Since these classes included orientation and course key concepts, not only climate and self-motivation, but also knowledge organization and mastery level were clearly insufficient for the knowledge receiving side.

Moreover, the late start resulted in uncertainty of the expectancies of both sides, and of their managers. The initial expectancy of the knowledge receiving side was that some of the learning goals were flexible and adjustable "on the fly" [46, 47]. However, the knowledge transmitting side's attitude appeared to be clearly different.

Moreover, in view of the knowledge transmitting side, the critical training processes were supposed to be prearranged and approved as training plans, otherwise certification goals would likely be missed. This resulted in an initial conceptual misunderstanding and a local knowledge transfer crisis.

To manage the crisis, both sides changed their processes for more flexible ones to the best extent possible. For a certain critical period of the preparatory activities, this resulted in a more complex training environment with frequent context switches, and increased the cognitive load for both sides. However, eventually the crisis was conquered, the training objectives met, and the trainees certified. This confirms that the wise application of the key human factors, irrespective of any cross-cultural issues and variations in expectancies and maturity levels, may result in conquering the local crises.

7   Crisis Agility: Enterprise Patterns Revisited

Crisis management of large-scale software products is a key problem in software and system engineering [45]. In this respect, one of the mission-critical issues is real-time system agility. The section discusses high-level architectural patterns and provides an outline of "crisis-proof" product development and process management. Within the software product architecture, we identify the agile objects, which are mission-critical in crisis. We introduce a process knowledge base and justify its role in the agility [52, 53].

7.1   Architecture Patterns: Enterprise Engineering Matrix

We introduce the architectural patterns in terms of the Enterprise Engineering Matrix; it includes processes, data and systems (see Fig. 12) [49–51].

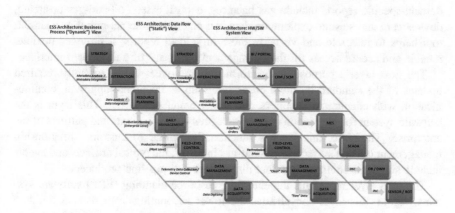

Fig. 12 The enterprise engineering matrix

The first, dynamic perspective shows the processes; it decomposes strategic goals into business processes, actions and tasks. The second, static perspective embraces the decomposed data objects used in the processes. The third, system perspective includes the large-scale systems that operate those data.

7.2 Process Management: Layer-Based Approach

In crisis, architectural agility becomes a concept that embraces the system qualities of "flexibility, balance, adaptability, and coordination" [31]. Architectural agility includes communication between the process participants and orchestration of the enhanced services. These efforts require collaboration with the customer and expert communities; they require CASE tool support in order to establish a common vision between the client and the developer [32].

In the 2000s, knowledge management used static database structures, such as wiki, blogs and content management systems. However, current semantic-based technologies of content and knowledge management moved to the process-level in order to increase knowledge availability.

Figure 12 outlines a high-level pattern of enterprise hardware-and-software systems [6, 33]. Our idea is that abstraction level, data aggregation degree and ability for strategic analysis, justification and decision-making grow bottom-up. Each layer communicates directly with the adjacent ones; it is the data/process consumer for lower levels and the provider for higher levels. Case study-based examples include nuclear power plant (NPP) and oil-and-gas sector.

The top, strategic layer of the enterprise software-and-hardware system is represented by the software toolkit for integrated representation of strategic management data. It is the dashboard, which top managers use to monitor dynamics of the key performance indicators. This dashboard aggregates data from the software systems for planning enterprise resources. General-purpose resources include human resources, financials, and time management. In oil-and-gas area, examples of domain-specific reports include gas balances, deposit assets, oil-and-gas upstream, downstream and seismic exploration data. An enterprise internet portal or a similar tool helps to integrate and visualize these high-level reports; it provides flexible, reliable and secure access for the management by means of a dashboard interface.

The next layer's purpose is to inform the employees of the urgently required updates of the standard business processes, such as document approval, communication with clients and suppliers, and target email messaging. This layer is the software system for interaction of the employees with the clients and partners of the enterprise. Therefore, this layer is functionally similar to customer relationship management (CRM) system. In oil-and-gas industry, the typical artifacts and agents include shipment contracts, product distributors and pipeline producers.

The next layer represents the enterprise resource planning (ERP) software systems. Again, this layer consolidates the lower accounting layer data; it builds a strategic view of the key business indicators. This level includes ERP software modules and subsystems, which assist in management and planning of oil-and-gas products, such as fixed assets, human and financial resources, documentation, supplies for oil-and-gas deposit construction and processing, supplies for NPP unit construction and electricity generation. Possible instances of such systems for the NPP industry include Siemens and Catie software products [34].

In contrast to the previous layers, the accounting layer contains lower-level software systems. While the resource layer focuses on analysis and forecasting of the key production indicators, such as revenues, profits, overheads, expenditures etc., the accounting layer's tasks are operational. The accounting layer includes software systems for warehousing, inventory management and the like. In NPP, the accounting layer includes the data on reaction unit construction, shipment and assembly, monitoring unit assembly maps, and technical conditions.

The next, supervisory or "drivers" layer, contains the software systems, which implement the interfaces between the software and the hardware components. This layer contains the SCADA systems for the end-users who interact with field-based devices and sensors, which perform plant operations, such as assembly-line production. In oil-and-gas, this supervisory layer typically deals with exploration and seismic data maps and the systems interacting with the devices and sensors, which perform plant operations, such as drilling exploration wells. In NPP, the supervisory layer typically deals with unit assembly maps and technical conditions; it interacts with the devices and sensors, which perform plant operations, such as power generation and reaction unit control.

The next, data layer embraces the databases and data warehouses. This data layer includes DBMS with data mining plug-ins, online analytical and transaction processing, middleware, and enterprise content management tools.

In the case of software-and-hardware system, essential features of the data layer are big data size, high availability, and heterogeneity. Heterogeneity has architectural and structural aspects [35]. Architectural heterogeneity includes non-normalized and legacy systems data; structural heterogeneity deals with weak-structured flows of audio and video data, and scanned documents. For instance, NPP requires integration with domain-specific applications for lifecycle management, electricity production and distribution, 6D modeling, and data visualization.

We can identify one more layer below the data layer; this is the hardware layer, which includes human-machine interfaces and such devices as programmable logic controllers, sensors and robots. The hardware layer aggregates the analog data, converts it into digital form, stores it and uses it for the higher-level enterprise applications.

Figure 13 presents the layout of the high-level pattern for the enterprise hardware-and-software systems. Let us discuss certain layers in more detail.

Figure 13 omits the relationships for the sake of space; however, it is clear that Layer 7 aggregates and consolidates the data objects of Layer 6 in order to provide a strategic view of the key business indicators.

"Dashboard" of key business indicators (portal)						7 Strategy
Supply Chain Management	Channel Management (Fax, email)	CRM	Campaigns	Marketing Plans		6 Relations
Accounts Payable	Accounts Receivable	General Ledger	Manufacturing	HR	PLM	5 Resource Management
Warehouse		Payroll		Inventory		4 Accounting
SCADA			CAD	CAM	CAE	3 "Drivers"
Databases / Data warehouses						2 Data
Hardware			Goods / Parts			1 Hardware

Fig. 13 Architectural pattern of the hardware-and-software systems

The Layer 6:

- Manages the channels of interaction with the clients, such as email, fax and IP telephony;
- Plans and manages events, such as special offers and sales;
- Manages distribution channels;
- Manages data on partners and clients, such as VIP clients, marketing campaigns and preferred communication methods.

Layer 6 consolidates Layer 5 data in order to represent an aggregated view of the key business indicators. Layer 5 includes the ERP components, which assist end-users in management and planning of fixed assets, payables, receivables, production and services, human resources, financials and documentation.

Compared to Layers 5–7, Layer 4 contains operational software systems for daily management.

Layer 3 contains the systems for the end-users who develop and manage design documentation, and who interact with field-based devices and sensors, which perform plant operations.

Layer 2 is a dedicated data layer, which could be drawn either to the side of the above five (Layers 3–7) representing their interactive penetration, or below them, limiting the interaction to Layer 3 only. The data level (Layer 2) is represented by databases and data warehouses; it contains system level software.

The only level that we identify below the data level is Layer 1, or the hardware level.

7.3 Layer-Based Approach Instance 1: Oil-and-Gas Corporation

Let us illustrate the above general pattern of the architectural layers by an example of the systems set used in the production and distribution activities of an oil-and-gas enterprise. Let us focus on the activities of a vertically integrated enterprise, which does exploration, production, processing and transportation of the oil-and-gas products.

In this case, Layer 7 is the enterprise internet portal, which allows enterprise managers to monitor the key performance indicators. The portal aggregates data and visually represents it in terms of high-level reports, including general and domain-specific ones, such as gas balances, deposit assets, upstream, downstream and seismic exploration data views.

Layer 6 informs employees of the urgently required updates of the standard business processes, such as approval of oil-and-gas shipment contracts, communication with oil-and-gas distributors and gas pipeline producers, and target email messaging of price lists for the produced and processed items. Layer 6 functions are: management of the employee interaction channels by means of gas communication, email, fax, intelligent IP telephony etc., planning and management of distribution networks, and data management for partners and clients, such as regular customers and VIP clients.

Layer 5 includes the ERP subsystems, which assist in management and planning of oil-and-gas production including fixed assets, supplies for deposit construction and oil/gas processing, human and financial resources, and documentation management.

Layer 4 contains the software systems for payroll and product supply management, including oil-and-gas transportation by pipelines, railways and ships, and the like.

Layer 3 contains the software systems for developers and managers of the design documentation including exploration and seismic data maps; it also includes the systems interacting with the devices and sensors, which perform plant operations, such as drilling exploration wells and oil-and-gas production.

Layer 2 often includes reliable and fault-tolerant Oracle DBMS-based solutions integrated with domain-specific ERP applications for upstream and downstream management and with online data visualization tools. The other reasons for using Oracle-based implementations are big data size and high availability.

Hardware Layer 1 includes the programming logic controllers and sensors used for exploratory drilling and oil-and-gas production; it operates with the analog data.

7.4   Layer-Based Approach Instance 2: Nuclear Power Plant

Let us illustrate the architectural pattern by an example of the systems set used in the production and distribution businesses of a nuclear power enterprise.

Layer 7 is the internet portal; it aggregates data for resource planning systems, including general and domain-specific ones, such as assembly maps, technical documentation, NPP monitoring data, production and distribution of the electricity, nuclear fuel supplies and waste utilization.

Layer 6 supports interaction of the key departments and employees of the NPP with their clients and partners. Its functions include informing the employees of the urgently required updates of their standard business processes, such as design documentation approval, communication with reaction unit customers and producers, target email messaging of the product articles and price lists for the produced NPP units.

Layer 5 represents the ERP and PLM, the systems for resource planning and production lifecycle management. It includes the modules, which help the managers to monitor and plan the production of NPP reaction units and their components. Possible instances of such systems for the NPP industry include Siemens and Catie software products [34, 35].

Layer 4 contains the software systems for product supply management, which monitor NPP payroll, human resource management, reaction unit construction, shipment and assembly, assembly maps for monitoring NPP units and technical conditions.

Layer 3 contains the software systems for developers and managers of the design documentation, such as NPP unit assembly maps, technical conditions, and the systems interacting with devices and sensors, which perform plant operations, such as heat generation, and which control reaction unit operating parameters, such as temperature and pressure.

Layer 2 typically uses Oracle DBMS-based solutions, which are custom-integrated with domain-specific PLM and ERP applications, and with 6D modeling and data visualization tools. The 6D models include 3D visualization of the NPP units and domain-specific models for the resources required to design and construct NPP, such as time, human resources and financials.

Layer 1 includes hardware devices, such as NPP programming logic controllers and sensors; it processes the analog data.

7.5   Process Knowledgebase: Adding Crisis Agility

As shown above, large clients typically have complex business processes and use a wide range of software products. Each of their products introduces customer-specific datasets that usually have a unique format; however, they can semantically intersect the adjacent datasets. To automate top-down and bottom-up process and data integration, large-scale clients require special-purpose software tools. Enterprise-scale integration bus is a sound basis for such tools; it requires a number of integration processes for management and maintenance.

In crisis, we suggest a centralized pattern-based process knowledgebase to consolidate, store and handle the standard processes and to provide typical problem solving methods.

The approach simplifies typical problem solving; it increases the software development productivity by means of reuse and cost reduction based on optimized decision-making.

The approach includes three key components:

- Basic processes, which include reference models for typical activities, inputs and outputs, and optional additional metrics;
- Business goals processor; it generates processes, which consist of basic sub-processes and allow achieving the goal(s) that the user requested;
- Integration module; it automates the processes by associating their steps with the data transmission and reception activities including stepwise notifications, acknowledgements and invitations.

The suggested approach is a key to crisis agility; it offers a repository of formal specifications and their implementations. It helps to handle uncertainty, incomplete documentation and requirement volatility. It is useful in top-down and bottom-up process construction. It assists in building processes based on formal specifications by means of composition of the existing artifacts, or by consecutive decomposition [36].

8   Conclusion

The chapter discussed crisis-oriented patterns and practices for software development and knowledge transfer. We used a set of models to represent and manage the transfer. For the basis of the models, we used informing science and a set of learning principles. We approached the software architecture in terms of process, data and system perspectives.

Our case studies included knowledge transfer from the Carnegie Mellon University masters' program in software engineering to the new and ambitious Russian Innopolis University.

The CMU MSIT/SE curriculum courses are tightly interwoven and integrated with project-based practices of software development. Their focus is quality and process improvement.

The knowledge transfer requires special instructor training in teaching proficiency. One example of such training is the Teaching Excellence Program (TEP) courses from iCarnegie Global Learning, a CMU subsidiary. Some of the learning principles we used were derived from these TEP courses.

We identified the following key factors, which may inhibit technology transfer:

- cross-cultural differences, such as mentality and language issues;
- geographical differences, such as time lags and vacation schedules;
- process maturity differences, for instance ad hoc versus defined and/or managed processes.

To promote technology transfer in crisis, we produced the following recommendations. These included multiple contexts and scaffolding, for instance: comparative learning of software development paradigms, environments, languages and tools; learning by doing versus formal models, and some other examples.

We recommend to minimize overall cognitive teaching-and-learning load. This includes faculty training plans with clearly separated concerns, allowing enough time for learning, environment preparation, and a few other aspects.

We consider efficient feedback a mission-critical activity to manage the crisis; to succeed, it should be frequent, timely, directed and goal-oriented.

In software engineering, "soft" skills that lead to required mastery level are critically important; these include communication, time management, negotiations, to name a few.

For the academic staff, the above recommendations should result in crisis-efficient metacognitive self-directed learning. This includes self-reflection, self-assessment and self-adjustment based on certain evidences, such as reflective reports, course improvement suggestions, additional lecture proposals and elective development.

The learning objectives should address higher levels of Bloom's taxonomy [37] including justification, analysis and practical application of the theory and principles learned. As a result, the students learn to act as software engineering professionals and to use their own expert-level judgment.

The above-mentioned multi-context education should include not only software development, but also its integration with hands-on learning-by-doing practices, including hardware interfaces. A number of "wicked" problems may happen when performing hands-on real-world tasks of a software engineer; these problems stimulate students to cope with the crisis conditions. However, the crisis is hardly manageable with underdeveloped "soft" skills and personal agility, which includes self-assessment, self-justification and self-adjustment. Even a project failure (likely to happen in a local crisis) is a starting point to learn, to adjust and to become more agile. This multi-disciplinary alloy of skills is required in the rapidly changing world of practical software engineering; it may become a remedy for crisis.

The model we used for the knowledge transfer included an informing framework, which involved transmitting and receiving sides, and the environment. In crisis, however, we would recommend certain adjustments and "add-ons" to this basic framework. First, to model resonant communication, we added the "amplifier" to the initial oscillating LC-type of the Cohen's model [1]. Second, to simulate feedback, which is a mission-critical crisis mechanism, we added positive and negative feedback paths for both transmitting and receiving sides. Third, we set the upper and lower thresholds for the informing amplified circuit in order to maintain a controlled resonance.

Our enhanced knowledge transfer model required compensation for the energy loss during the transfer due to physical nature of the circuit. The reasons of this loss included cross-cultural, inter- and intra-organizational barriers. The feedback should be negative in order to balance the internal resonance between the inter- and intra-organizational parts of the informing system. In crisis, investing excessive energy (i.e. resources) into the informing system would likely result in its crash and even physical collapse due to uncontrolled resonance. Therefore, we should control the value of resonance by a bidirectional, positive and negative, feedback. In case the energy increase received from the feedback exceeds a certain threshold, a negative impulse is required to compensate that. In a number of crisis scenarios, it is practically sufficient to stop adding energy to the informing system (or at least to its certain part) until it reaches the lower energy threshold, where the (sub)system already starts to slow down, even though the feedback value might still remain positive.

In crisis, knowledge transfer typically requires special training of the receiving side, i.e. the faculty. That is why the lessons learned at CMU while doing MSIT/SE faculty training were mission-critical. These lessons included the following. We identified the essential human-related factors for successful knowledge transfer: prior knowledge, knowledge organization, feedback, mastery and practice. Of these, the first three factors were mission-critical.

In crisis, an adequate, bidirectional feedback-driven meta-cognitive cycle organization was very important to ensure the quality of learning. That is why the intended learning objectives should address the entire meta-cognitive cycle, including such activities as reflection, analysis, justification, adjustment and practical application of the theory and principles learned.

Therewith, both the students and the faculty should act as software engineering professionals and use their own expert-level judgment. Knowledge transfer for software engineering should be multi-context; it should include not only theoretical software development but also hands-on real-world project practice. Teamwork, communication and time management are the essential "soft" skills. These multi-disciplinary skills are mission-critical in a crisis, with rapidly changing and complex environments. In a crisis, even a project failure may become a possible starting point to reflect, adjust and become agile.

After analyzing the knowledge transfer models based on the informing framework, learning principles and practices, we discussed an approach for representing enterprise software systems based on high-level design patterns. We suggested a general architectural framework for such systems-of-systems, which included five application levels and two data levels. The application levels ranged from strategic decision-making and key interface components down to "drivers". The data levels included the "processed" digital and the "raw" analog data. Finally, we instantiated this high-level design pattern by the functional outlines for systems-of-systems in oil-and-gas and nuclear power industries.

References

1. Cohen, E.B.: Reconceptualizing information systems as a field of the transdiscipline informing science: from ugly duckling to Swan. J. Comput. Inf. Technol. 7(3), 213–219
2. Ambrose, S.A., Bridges, M.W., DiPietro, M., Lovett, M.C., Norman, M.K.: How Learning Works. Seven Research-Based Principles for Smart Teaching. John Wiley & Sons (2010)
3. Shannon, C.E., Weaver, W.: The mathematical theory of communication. University of Illinois Press, Urbana (1949)
4. Gill, G., Bhattacherjee, A.: The informing sciences at a crossroads: the role of the client. Informing Sci. 10(17), (2007)
5. Lattanze, A.: Architecting Software Intensive Systems: A Practitioner's Guide. Auerbach (2008)
6. Hohpe, G., Woolf, B.: Enterprise Integration Patterns: Designing, Building, and Deploying Messaging Solutions. Addison-Wesley (2004)
7. CMMI Standard. http://www.sei.cmu.edu/cmmi/. Retrieved 25 Nov 2015
8. Kuchins, A.C., Beavin, A., Bryndza, A.: Russia's 2020 Strategic Economic Goals and the Role of International Integration. Center for Strategic and International Studies, Washington, D.C. (2008)
9. Kondratiev, D., Tormasov, A., Stanko, T., Jones, R., Taran G.: Innopolis University—a new IT resource for Russia. In: Proceedings of the International Conference on Interactive Collaborative Learning (ICL), Kazan, Russia (2013)
10. US News Colleges Rankings and Reviews. http://colleges.usnews.rankingsandreviews.com/best-colleges/carnegiemellon-university-3242. Retrieved 25 Nov 2015
11. Hattie, J., Timperley, H.: The power of feedback. Rev. Educ. Res. 77, 81–112 (2007)
12. Alvermann, D.E., Smith, L.C., Readence, J.E.: Prior knowledge activation and the comprehension of compatible and incompatible text. Read. Res. Q. 20(4), 420–436 (1985). (John Wiley & Sons)
13. Chi, M.T.H., Bassock, M., Lewis, M.W., Rainman, P., Glaser, R.: Self-explanations: how students study and use examples in learning to solve problems. Cogn. Sci. 13, 145–182 (1989)
14. Gick, M.L., Holyoak, K.J.: Analogical problem solving. Cogn. Psychol. 12, 306–355 (1980)
15. Ames, C.: Motivation: what teachers need to know. Teachers Coll. Rec. 91, 409–472 (1990)
16. McGregor, H., Elliot, A.: Achievement goals as predictors of achievement—relevant processes prior to task engagement. J. Educ. Psychol. 94, 381–395 (2002)
17. Barron, K., Harackiewicz, J.: Achievement goals and optimal motivation: testing multiple goal models. J. Pers. Soc. Psychol. 80, 706–722 (2001)
18. National Research Council How people learn: Brain, mind, experience, and school. National Academy Press, Washington, D.C. (2000)
19. Bishop, C.A.: Case studies in systems engineering—central to the success of applied systems engineering education program. In: (IMCIC-2012), Orlando, FL, USA (2012)
20. Carlile, O., Jordan, A.: It works in practice but will it work in theory? The theoretical underpinnings of pedagogy. In: Emerging issues in the Practice of University Learning and Teaching, pp. 11–26. All Ireland Society for Higher Education, Dublin (2005)

21. Gentner, D., Loewenstein, J., Thompson, L.: Learning and transfer: a general role for analogical encoding. J. Educ. Psychol. **95**, 393–405 (2003)
22. Cognition and Technology Group at Vanderbilt From visual word problems to learning communities: changing conceptions of cognitive research. In K. McGilly (ed.) Classroom Lessons: Integrating Cognitive Theory and Classroom Practice, pp. 157–200. MIT Press/Bradford Books, Cambridge, MA (1994)
23. Schwartz, D.L., Bransford, J.D.: A time for telling. Cogn. Instr. **16**, 475–522 (1998)
24. Novak, J.: Learning, Creating, and using Knowledge: Concept Maps as Facilitative Tools in Schools and Corporations. Erlbaum, Mahwah, NJ (1998)
25. Innopolis University Presentation. http://innopolis.ru/files/docs/uni/innopolis_university.pdf. Retrieved 25 Nov 2015
26. Humpfrey, W.S.: A Discipline for Software Engineering. Addison Wesley (1995)
27. The Software Process Dashboard Initiative The Software Process Dashboard Initiative. http://www.processdash.com/. Retrieved 25 Nov 2015
28. Bass, L., Clements, P., Kazman, R.: Software architecture in practice, 3d edn., 589 p. Pearson Education, Inc. (2012)
29. Stevens, D.D., Levi, A.J.: Introduction to rubrics: an assessment tool to save grading time, convey effective feedback and promote student learning. Stylus, Sterling, VA (2005)
30. Biggs, J.: Aligning teaching for constructing learning. High. Educ. Acad. Asset (2004)
31. Dyer, L., Ericksen, J.: Complexity-based agile enterprises: putting self-organizing emergence to work. In: Wilkinson, A., et al. (eds.) The Sage Handbook of Human Resource Management, pp. 436–457. Sage, London (2009)
32. Gromoff, A., Kazantsev, N., Kozhevnikov, D., Ponfilenok, M., Stavenko, Y. Newer approach to create flexible business architecture of modern enterprise. Global J. Flex. Syst. Manage. **13** (4), 207–215 (2012). (Springer-Verlag)
33. Fowler, M.: Patterns of Enterprise Application Architecture. Addison-Wesley (2002)
34. Zykov, S.: Pattern development technology for heterogeneous enterprise software systems. J. Commun. Comput. **7**(4), 56–61 (2010)
35. Zykov, S.V.: Architecturing software engineering ecosystem. In: Proceedings of the e-Skills for Knowledge Production and Innovation Conference, pp. 543–550. Cape Town, South Africa (2014)
36. Fleischmann, A.: What Is S-BPM? S-BPM ONE—Setting the Stage for Subject-Oriented Business Process Management. Communications in Computer and Information Science. Springer, Berlin, Heidelberg (2010)
37. Bloom, B.S.: Taxonomy of educational objectives. The classification of educational goals. Longmans, London (1964)
38. Anderson, J.R., Corbett, A.T., Koedinger, K.R., Pelletier, R.: Cognitive tutors: lessons learned. J. Learn. Sci. **4**, 167–207 (1995)
39. Laurillard, D.: Chapter 10: designing teaching materials. Rethinking university teaching: a framework for the effective use of learning technologies, pp. 181–198. Routledge, London (2002)
40. Pascarella, E., Terenzini, P.: How college affects students: a third decade of research. Jossey-Bass, San Francisco (2005)
41. Resnick, L.B.: Mathematics and science learning. Science **220**, 477–478 (1983)
42. Sprague, J., Stuart, D.: The speaker's handbook. Harcourt College Publishers, Fort Worth, TX (2000)
43. The Carnegie Mellon University History. http://www.cmu.edu/about/history. Retrieved 25 Nov 2015
44. The Carnegie Mellon University. http://www.cs.cmu.edu/. Retrieved 25 Nov 2015
45. The First NATO Conference on Software Engineering. dl.acm.org/citation.cfm?id=1102020. Retrieved 25 Nov 2015
46. Waters, D.J., Waters, L.S.: On the self-renewal of teachers. JVME **38**(3), 235–241 (2011)
47. Wigfield, A., Eccles, J.: Expectancy-value theory of achievement motivation. Contemp. Educ. Psychol. **25**, 68–81 (2000)

48. Zimmerman, B.J.: Theories of self-regulated learning and academic achievement: an overview and analysis. In: Zimmerman, B.J., Schunk, D.H. (eds.) self-regulated learning and academic achievement, 2nd edn, pp. 1–3. Erlbaum, Hillsdale, NJ (2001)
49. Gamma, E., Helm, R., Johnson, R., Vlissides, J.: Design Patterns CD: Elements of Reusable Object-Oriented Software. Addison-Wesley (1998)
50. Freeman, E., Bates, B., Sierra, K., Robson, E.: Head First Design Patterns. O'Reilly (2004)
51. Zykov, S.: Designing patterns to support heterogeneous enterprise systems lifecycle. In: Proceedings of the 5th Central and Eastern European Software Engineering Conference in Russia (CEE-SECR) (2009)
52. Sørensen, M.H., Urzyczyn, P.: Lectures on the Curry-Howard Isomorphism, vol. 149. Elsevier (2006)
53. Baader, F., et al.: The Description Logic Handbook: Theory, Implementation, and Applications. Cambridge University, Cambridge (2007)
54. CMMI® for Development, Version 1.3. http://www.sei.cmu.edu/reports/10tr033.pdf. Retrieved 25 Nov 2015

Conclusion. Can We Manage the Crisis?

We have discussed lifecycle optimization of software production for crisis management by means of software engineering methods and tools.

Our outcomes are based on the lessons learned from the software engineering crisis which started in the 1960s. At that time, software product lifecycles, which the industry had just started moving toward, were anarchic in many ways, since no systematic approach existed. However, so far there has been no single answer whether the crisis is over; some researchers argue that it has been conquered, while the others say it is still here.

We start from analyzing the findings of Marx, a pioneering researcher of the economic crises. In his terms, a typical reason for any crisis is the "anarchy of production", which results from the absence of centralized planning and regulation of the production and its lifecycle. He also stated that crises result from a misbalanced production and the realization of a surplus value, the root cause of which is the separation between the producers and the means of production. Disproportionate investment and over-expansion of productive capacity often trigger a crisis first for the enterprise, and then on industrial scale. For the IT industry, complexity and rapid growth add more probability for crises to hit software production.

We conclude that the crisis might result from inadequate planning and from a lack of common understanding of the product size and scope between the software producers and the software consumers. In our opinion, the root cause of the crisis is the resource misbalance due to an inadequate, inappropriate or distorted common vision of the product features and/or project constraints between the client and the developer. Clearly, each of these two sides has a number of participants with very different preferences, goals and expectations. Thus, the crises in software engineering depend not only on technology-related but also on human-related factors. In other words, human nature is an essential source of the crises in software production; however, any crisis manifests itself in software project economics.

Since the crisis hydra has at least two heads, one of which is technological, and the other has a human-related nature, we select a complex approach to conquer it.

Thus, we propose an adaptive methodology for software product development, which allows avoiding local crises of software production. The idea is to optimize

© Springer International Publishing Switzerland 2016   119
S.V. Zykov, *Crisis Management for Software Development*
and Knowledge Transfer, Smart Innovation, Systems and Technologies 61,
DOI 10.1007/978-3-319-42966-3

the software product lifecycle. A typical methodology includes models, methods and tools. Following this approach, we start from the high-level models for software development lifecycles; each of the models sequentially elaborates the software product from its initial idea to a utilizable implementation.

We discuss a general lifecycle pattern and its stages, and analyze their impact on the timeframe and budget of software product development. We find that the maintenance stage is critically important as it often contributes to nearly 70 % of the project expenses; poor maintainability is a likely source of crisis. Another recommendation is to reduce the cognitive load of maintenance so that the developers can sequentially apply each type of maintenance one after another. We find that the cost of defect detection and fixing increases exponentially as we move from the earlier lifecycle stages to the later ones, so defect detection in crisis should start as early as possible.

Concerning the lifecycle models, some of these require iterative development, whereas the others are more straightforward. Certain models require a relatively high level of discipline and organizational maturity; otherwise, the lifecycle could easily degenerate into a trial and error approach. We find that the model selection influences a number of critical parameters of the software development project, such as architecture, budget and timeframe, and it often determines its overall success. The lifecycle model selection should be adequate to the experience of the project team in terms of problem domain expertise and operational knowledge of the required technologies, tools and standards. The model selection also determines the product artifacts and quality attributes; it helps to decrease the time to market, which is mission-critical in a crisis. Software engineering uses product quality metrics, which makes project monitoring and management more accurate and predictable in a crisis.

We identify the key advantages and disadvantages for each model discussed; we conclude that there is no "silver bullet", or no universal model, which would suit all software production equally well. We find that the scope and size of the project determines lifecycle model selection, and we recommend customizing and combining the models in order to adjust to the specific features of the product.

We discuss software development methodologies. These are adaptive process frameworks, adjustable to software product size and scope. Each methodology includes a set of methods, principles and techniques, and software development tools.

Each of the methodologies that we discuss is flexible enough to implement any of these lifecycle models. Some methodologies are more formal, others are more agile. In crisis conditions, such as hard-to-formalize problems, rapidly changing requirements and other uncertainties, agile methodologies, which usually have fewer artifacts, are applicable. However, agile methodologies require extremely well disciplined development, and, consequently, they impose extra human factor-related constraints.

No surprise that we find no "silver bullet" for software development methodologies. However, due to rigorous processes and well-defined deliverables, formal methodologies are better for mission-critical and large-scale applications;

crisis-agile methodologies might result in low quality software products and build-and-fix lifecycles in an undisciplined development.

Following the sequential elaboration strategy, after the lifecycle models and development methodologies we move on to the patterns and practices of resource efficient software production; this is mission-critical in a crisis. We discuss a methodology of pattern-based software product development, which includes a set of formal models, processes, methods and tools.

To make the product development crisis-agile, we suggest a resource-efficient component approach based on high-level architecture patterns for frequently used combinations of large-scale product modules. We use these combinations as baselines and branches to manage crisis-related requirement changes, frequent release updates and typical requests from different clients. We recommend to support this pattern-based development by means of domain-specific languages and visual computer-aided tools.

We also propose a specific incremental software development methodology for crisis-agile software development of large-scale, distributed heterogeneous applications.

These methodologies provide an industrial level of software quality; they efficiently reduce project terms and costs for large-scale heterogeneous products, such as governmental and commercial enterprise software applications. Implementation areas include oil-and-gas-production, air transportation, retail networks and nuclear power generation. For each implementation, we develop a domain-specific language, which helps in pattern-based product re-engineering, cloning, adjustment and maintenance. We give crisis-agile recommendations for the application of the approach to existing clients and to new customers.

We also suggest an enhanced architectural pattern for systems-of-systems, which includes five application levels and two data levels. The application levels range from strategic decision-making and key interface components down to "drivers". The data levels include the "processed" digital and the "raw" analog data. We instantiate the high-level design pattern by the functional outlines for systems-of-systems in the oil-and-gas and nuclear power industries.

The next level of crisis-oriented patterns and practices is intended for software development and knowledge transfer. We use a set of models to represent and manage the transfer. We use informing science models and a set of specific learning principles. We approach software architecture in terms of process, data and system perspectives. Our case studies include knowledge transfer of the masters' program in software engineering from the Carnegie Mellon University to the Russian Innopolis University. The CMU curriculum is tightly interwoven and integrated with a project-based practice of software development; its focus is on quality and process improvement.

We identify the key factors, which might inhibit technology transfer; these include differences in culture, geography and process maturity.

To promote crisis technology transfer, our recommendations include multiple contexts, scaffolding and learning by doing. Other recommendations are to reduce the cognitive teaching-and-learning load, to establish and maintain frequent, timely, directed and goal-oriented feedback, and to build up personal crisis-agile "soft" skills, such as communication, time management, negotiations, self-reflection, self-assessment and self-adjustment. We also recommend addressing higher levels of Bloom's taxonomy including justification, analysis and the practical application of the theory and principles learned. We find that this multi-disciplinary alloy of knowledge, skills and attitudes should be a remedy in crisis.

For knowledge transfer, we also apply the model, which includes the informing framework represented by the transmitting and the receiving sides, and the environment. To model resonant communication and feedback, we enhance crisis agility of this basic model framework by an amplified circuit with controlled resonance and a bidirectional feedback loop for both sides.

We find that in crisis, knowledge transfer requires special training of the receiving side based on such essential human-related factors as prior knowledge, knowledge organization, feedback, mastery and practice; of these, the first three are mission-critical. A bidirectional feedback-driven meta-cognitive cycle is critically important for the quality of learning. We recommend multi-context knowledge transfer, which includes hands-on real-world project practice, and "soft" skills.

Thus, we use a multi-faceted approach to software engineering and knowledge transfer, which includes human and technological factors. The systematic approach we use includes formal models, a set of domain-specific methods and visual tools; it increases crisis agility so that software development becomes more predictable, accurate and adaptive at the same time.

In our view, the root cause of the software development crisis is the human mind itself, and we can manage the crisis if we approach human-related and technology-related issues and challenges in a systematic and disciplined way.

Glossary

Activity A thing that a person or group does or has done, a relatively small isolated task with clear exit criteria

Adaptive maintenance Maintenance, which modifies system in order to adapt the product to the new software and hardware environment

Agile methodology An alternative to traditional project management where emphasis is placed on empowering people to collaborate and make team decisions in addition to continuous planning, continuous testing and continuous integration

Agility Ability to adapt to uncertainties and changes of environment

Anthropic-oriented Relating to human

Architectural heterogeneity A property of a system, which includes components based on different architectural patterns

Architecture Centric Design Method (ACDM) A novel method for software architectural design developed by the Software Engineering Institute at the Carnegie Mellon University

Architecture Tradeoff Analysis Method (ATAM) Risk-mitigation process used early in the software development life cycle. ATAM was developed by the Software Engineering Institute at the Carnegie Mellon University

Backlog An accumulation of uncompleted work or matters needing to be dealt with

Best practice(s) Commercial or professional procedures that are accepted or prescribed as being correct or most effective

Bloom's taxonomy Bloom's classification of levels of intellectual behavior important in learning

Build-and-fix, model A model of software development without a deliberate strategy or methodology. Programmers immediately begin coding. Often late in

© Springer International Publishing Switzerland 2016
S.V. Zykov, *Crisis Management for Software Development
and Knowledge Transfer*, Smart Innovation, Systems and Technologies 61,
DOI 10.1007/978-3-319-42966-3

the development cycle, testing begins, and the unavoidable bugs must then be fixed before the product delivery

Business requirements Critical activities of an enterprise that must be performed to meet the organizational objective(s) while remaining solution independent

Cohesion Action or fact of forming a united whole

Common vision Essential component of a learning organization because it provides the focus and energy for learning; a realistic, credible, attractive future for an organization Common vision is derived from the members of the organization, creating common interests and a sense of shared purpose for all organizational activities

Computer-aided design (CAD) Software used by architects, engineers, drafters, artists, and others to create precision drawings or technical illustrations. CAD software can be used to create two-dimensional (2D) drawings or three-dimensional (3D) models

Computer-aided engineering (CAE) Broad usage of computer software to aid in engineering analysis tasks

Computer-aided manufacturing (CAM) An application technology that uses computer software and machinery to facilitate and automate manufacturing processes

Corrective maintenance Maintenance, which fixes existing defects in the software product without changing the design specifications

Coupling The pairing of two items

Courage The ability to do something that frightens one; bravery

Course climate A set of overarching and pervasive values, norms, relationships, and policies of the course that make its character

Crisis Misbalanced production and realization of a surplus value, the root cause of which is separation between the producers and the means of production

Customer relations management (CRM) Practices, strategies and technologies that companies use to manage and analyze customer interactions and data throughout the customer lifecycle, with the goal of improving business relationships with customers, assisting in customer retention and driving sales growth

Data warehouse A large store of data accumulated from a wide range of sources within a company and used to guide management decisions

Database management system (DBMS) Software that handles the storage, retrieval, and updating of data in a computer system

Deliverable A practical outcome, any measurable project artifact, which is a result of each project task, work item or activity

Design pattern A general reusable solution to a commonly occurring problem within a given context in software design

Design specification A detailed document providing information about the characteristics of a project to set criteria the developers will need to meet

Design A software development lifecycle stage, which formally describes the components of the future software product and the connections between these components

Enterprise content management (ECM) Formalized means of organizing and storing an organization's documents, and other content that relate to the organization's processes. Encompasses strategies, methods and tools used throughout the lifecycle of the content

Enterprise Engineering Matrix Matrix, the columns of which correspond to processes, data and systems, and the rows of which contain enterprise system levels. Used to detect and predict local crises of software production

Enterprise Resource Planning (ERP) Business process management software that allows an organization to use a system of integrated applications to manage the business and automate many back office functions related to technology, services and human resources

Evolutionary, model An iterative model of software development based on the idea of rapidly developing an initial software implementation from very abstract specifications and modifying this according to appraisal

Expectancy The feeling that a person has when he or she is expecting something

Extreme programming, methodology (XP) A software development methodology, which is intended to improve software quality and responsiveness to changing customer requirements; a pragmatic approach to program development that emphasizes business results first and takes an incremental approach to building the product through continuous testing and revision

Far knowledge transfer Knowledge transfer that allows a multi-domain knowledge application in different contexts, some of which are substantially far from the original one

Feedback A helpful information or criticism that is given to someone to say what can be done to improve performance

Formal methodology A codified set of practices (sometimes accompanied by training materials, formal educational programs, worksheets, and diagramming tools) that may be repeatedly carried out to produce software

"Fragile" base classes A problem of object-oriented programming where the superclasses contain seemingly safe modifications, which, when inherited by the derived classes, may cause malfunctions of these derived classes

Heterogeneity A property of a set, which consists of elements that are essentially different from each other

Human-Related Factor A factor originating from human nature, which influences requirements elicitation, and, consequently, software development

Implementation A software development lifecycle stage, which produces the code of each individual component of the software product

Incremental, model A model of software development where the product is designed, implemented and tested incrementally (a little more is added each time) until the product is finished. It involves both development and maintenance

Informing science A transdiscipline to promote the study of informing processes across a set of disciplines, including management information systems, education, business, instructional technology, computer science, communications, psychology, philosophy, library science and information science. The unit of analysis is the informing system, which is a collection of informers, clients and channels serving a particular informing need

In-house development Software development by a corporate entity for purpose of using it within the organization

Integration A software development lifecycle stage, which produces the entire software product out of the individual components implemented previously

Interrelatedness The ability of one system component change to significantly influence a number of adjacent components changes

Key performance indicator Business metric used to evaluate factors that are crucial to the success of an organization

Knowledge transfer (KT) The practical problem of transferring knowledge from one part of the organization to another

Legacy software product An old software product, of, relating to, or being a previous or outdated computer system. Often implies that the system is out of date or in need of replacement

Maintenance A software development lifecycle stage, which includes all aspects of the product operation and support at the client's site

Mastery A skill that allows doing, using, or understanding something very well

Metacognitive Something that refers to higher order thinking which involves active control over the cognitive processes engaged in learning

Meta-knowledge transfer Practical problem of transferring knowledge about a preselected knowledge

Metaphor A figure of speech in which a word or phrase is applied to an object or action to which it is not literally applicable

Method A particular procedure for accomplishing or approaching something, especially a systematic or established one

Methodology A system of methods used in a particular area of study or activity; a framework that is used to structure, plan and control the process of developing an information system

Milestone A key control point where certain results are achieved; a significant stage or event in the development of something

Modularity The degree to which a system's components may be separated and recombined; uses minimum connectivity between the modules, so that each relatively small and functionally separate task is located in a separate software module

Motivation A force or influence that causes someone to do something

MSF, methodology A set of principles, models, disciplines, concepts, and guidelines for delivering information technology solutions from Microsoft. MSF is not limited to developing applications only; it is also applicable to deployment, networking and infrastructure projects

Near knowledge transfer The knowledge transfer applicable to adjacent problem domains only

Object-oriented, model A model of software development based on object-oriented paradigm

Online analytical processing (OLAP) A category of software tools that provides analysis of data stored in a database. OLAP tools enable users to analyze different dimensions of multidimensional data

Online transaction processing (OLTP) A class of information systems that facilitate and manage transaction-oriented applications, typically for data entry and retrieval transaction processing

Oscillator (LC) circuit An electric circuit, which consists of capacitor and inductive coupling. Uses feedback for oscillation

Perfective maintenance A type of maintenance, which implements changes to the product functional specification, making the new product release with improved functionality and same or better quality in terms of performance, reliability, security, availability, usability etc.

Practice An activity of doing something repeatedly in order to become better at it

Process A series of actions or steps taken in order to achieve a goal; a sequence of the tasks to be implemented, they are clearly different, i.e. have a clear start and termination criteria, and, in some times dependent on each other

Production lifecycle management (PLM) Process of managing the entire lifecycle of a product from inception, through engineering design and manufacture, to service and disposal of manufactured products

Quality Attribute (QA) A systemic property of a software product, which critically influences its quality

Rapid prototyping, model The activity of creating prototypes of software applications, i.e. incomplete versions of the software being developed. A prototype typically simulates only a few aspects of, and may be completely different from, the final product

Refactoring The process of restructuring existing computer code without changing its external behavior. Refactoring improves nonfunctional attributes of the software

Requirements analysis A software development lifecycle stage, which identifies the desired properties of the future software product

Requirements specification A software development lifecycle stage, which formally describes the properties of the future software product

Resonance A quality of evoking response

Retirement A software development lifecycle stage, when the product is completely and permanently put out of operation

Return on investment (ROI) A performance measure used to evaluate the efficiency of an investment or to compare the efficiency of a number of different investments. ROI measures the amount of return on an investment relative to the investment's cost

RUP, methodology Rational Unified Process, a software development methodology from Rational. Based on UML language, RUP organizes the development of software into four phases, each consisting of one or more executable iterations of the software at that stage of development: inception, elaboration, construction, transition

Scaffolding A variety of instructional techniques used to move students progressively toward stronger understanding and, ultimately, greater independence in the learning process

Scalability The capability of a system, network, or process to handle a growing amount of work, or its potential to be enlarged in order to accommodate that growth

Scope, product scope Features and functions that characterize a product, service or result; product scope defines what the product will look like, how will it work, its features, etc.

Scrum master The facilitator for a product development team that uses scrum, a rugby analogy for a development methodology that allows a team to self-organize and make changes quickly. The scrum master manages the process for how information is exchanged

Scrum, methodology An iterative and incremental agile software development methodology for managing product development

Self-adjustment Adjustment of oneself or itself, as to the environment

Size, product size Overall size of the software being built or modified

"Soft" skills Personal attributes, which indicate a high level of emotional intelligence, such as teamwork, negotiations etc.

Software Engineering A set of tasks, methods, tools and technologies used to design and implement complex, replicable and high-quality software systems

Software environment Surroundings for an application; usually includes operating system, database system and development tools

Software process An over-arching process of developing a software product

Software product Merchandise consisting of a computer program that is offered for sale

Spiral, model Systems development lifecycle model, which combines the features of the prototyping model and the waterfall model

Sprint A set period of time during which specific work has to be completed and made ready for review

Stakeholder A person with an interest or concern in something

Structural heterogeneity A property of a dataset, which includes both strong and weak-structured elements

Synchronize and stabilize, model A systems development life cycle model in which teams work in parallel on individual application modules, frequently synchronizing their code with that of other teams, and regularly debugging, i.e. stabilizing the code

System-of-systems A viewing of multiple, dispersed, independent systems in context as part of a larger, more complex system

Technical constraint A technical limitation or restriction

Tool Computer-aided software, which supports the software development processes and methods; typically used for software development or system maintenance

Vision, product vision An original idea, clear yet informal, of the fundamental differences and customer values for the future software product as compared to the existing ones, and its benefits after the implementation; desired future state that would be achieved by developing and deploying a product

Waterfall, model Sequential design process, used in software development processes, in which progress is seen as flowing steadily downwards through the phases of conception, initiation, analysis, design, construction, testing, production/implementation and maintenance

Index

© Springer International Publishing Switzerland 2016
S.V. Zykov, *Crisis Management for Software Development
and Knowledge Transfer*, Smart Innovation, Systems and Technologies 61,
DOI 10.1007/978-3-319-42966-3